# HEROES OF THE SPACE AGE

# HEROES OF THE
# SPACE AGE

## INCREDIBLE STORIES OF THE FAMOUS AND FORGOTTEN
## MEN AND WOMEN WHO
## TOOK HUMANITY TO THE STARS

### ROD PYLE

Prometheus Books

59 John Glenn Drive
Amherst, New York 14228

Inquiries should be addressed to
Prometheus Books | 59 John Glenn Drive | Amherst, New York 14228
VOICE: 716–691–0133 • FAX: 716–691–0137
WWW.PROMETHEUSBOOKS.COM

23 22 21 20 19    5 4 3 2 1

Library of Congress Cataloging-in-Publication Data

Names: Pyle, Rod, author.
Title: Heroes of the space age : incredible stories of the famous and forgotten men
    and women who took humanity to the stars / by Rod Pyle.
Description: Amherst, New York : Prometheus Books, 2019.
Identifiers: LCCN 2018056213 (print) | LCCN 2018058090 (ebook) |
    ISBN 9781633885257 (ebook) | ISBN 9781633885240 (pbk.)
Subjects: LCSH: United States. National Aeronautics and Space Administration—
    Officials and employees—Biography. | Astronauts—United States—Biography
    | Astronauts—Soviet Union—Biography. | Gagarin, Yuri Alekseyevich, 1934-
    1968. | Glenn, John, 1921-2016. | Nikolaeva-Tereshkova, Valentina Vladimirovna,
    1937- | Kranz, Gene. | Hamilton, Margaret Heafield, 1936- | Armstrong, Neil,
    1930-2012. | Aldrin, Buzz. | Conrad, Pete, 1930-1999.
Classification: LCC TL789.85.A1 (ebook) | LCC TL789.85.A1 P945 2019 (print) |
    DDC 629.450092/2—dc23
LC record available at https://lccn.loc.gov/2018056213

Printed in the United States of America

*To Gloria Lum,*
*a wonderful companion who made it all possible.*

# CONTENTS

# INTRODUCTION

Writing a book about some of the outstanding figures from the space age has been a wonderful assignment. Choosing which of those amazing souls to profile has been a curse. There are so many inspiring people who were part of that great adventure that it's downright painful to pick only a handful of them for inclusion. Many readers will find their favorites within these pages, while others may discover that their most prized figure from that era is omitted. By way of advance apology, I will merely offer this: I selected some of the most amazing people I knew of, and I have tried to bring them to life in this book.

"Heroes" is a slippery word. A typical dictionary reference will define a hero as "A person who is admired or idealized for courage, outstanding achievements, or noble qualities." This definition does not include the flaws, limitations, or shortcomings any person has regardless of their greatness, and these are things that are to be expected—we are all only human. Still, flaws aside, each figure profiled in this book stood for something greater than themselves; in this case, the exploration of space. Each did so in different but equally remarkable ways. Some were probably better wives and husbands than others, some were doubtless more regular churchgoers or volunteered at their PTA more frequently than their peers, but all these folks contributed in some remarkable way to the incredible spirit of exploration that defined the years between 1957 and 1973, the rough boundaries of the space race.

If I have missed your favorite person from that era, please forgive me—such an omission was likely a case of editorial triage; it is certainly not a deliberate slight of any one of the many hundreds of gifted people who worked at the forefront of the space race, and who labored so tirelessly to reach for the stars. When writing about that golden age of space exploration, finding diversity within the figures profiled was a challenge—

there were far too few women and people of non-European origins to choose from. I'm happy to report, however, that this no longer holds true within NASA, though there is still much to be done to bring true diversity to spaceflight endeavors. Progress toward diversity and inclusion is being made, and that can only strengthen our collective, global efforts.

The space age was just the first act of a story we are seeing revived today, as private entrepreneurs, government officials, and countless engineers, scientists, and technicians strive to create the next act in space exploration—a new age that is unfolding as we watch. The opportunities that are opening up are truly inspiring, and countless people from all genders, ethnicities, and a diverse swath of cultural and national backgrounds are entering this grand adventure.

In another decade or so, when someone else prepares to write a book about the heroes of the second space age, they will have to make their own set of choices about whom they feature and whom they will have to omit, but whatever selections they make will be far more reflective of the vast and varied cultures that make up our world. With luck, they will be writing about men and women from many nations and a dozen ethnicities, who built rockets, founded outposts on other worlds, engendered scientific and technological breakthroughs, and went forth to settle the vast domain beyond our own fragile planet. In doing so they will have continued to improve and expand what makes our species great, will have improved life on the planet that has nurtured us for so long, and will have made space a new home for humanity. I look forward to reading that next volume in the greatest of human endeavors, the exploration of space.

Special thanks to Matthew T. Polomik for a thorough technical edit of this book. Also, my gratitude to James Hansen, Francis French, Colin Burgess, Nick Taylor, James Oberg, Adam Fabio, Matthew McConnel, Stephen Ambrose, Douglas Brinkley, Eric Jones, Ken Glover, Nancy Conrad, Howard Klausner, Mark Wade, Andrew Chaikin, the staff of NASA's historical section, and the myriad other authors who have so deftly and compellingly covered many of the dynamic personalities who populated the space age.

# CHAPTER 1

# YURI GAGARIN: THE FIRST STARMAN

What kind of man allows himself to be pushed, pulled, nearly suffocated, half-drowned, almost crushed, then jammed into a tiny metal sphere and hurled into space atop a rocket that has launched only small dogs before him (with a high rate of mortality)?

> Modest; embarrasses when his humor gets a little too racy; high degree of intellectual development evident in Yuri; fantastic memory; distinguishes himself from his colleagues by his sharp and far-ranging sense of attention to his surroundings; a well-developed imagination; quick reactions; persevering, prepares himself painstakingly for his activities and training exercises, handles celestial mechanics and mathematical formulae with ease as well as excels in higher mathematics; does not feel constrained when he has to defend his point of view if he considers himself right; appears that he understands life better than a lot of his friends.[1]

Thus was Yuri Gagarin, the first human in space, described by a Soviet Air Force doctor in 1960. This was at the tail end of a rigorous, intensive selection process, and the young, easily embarrassed pilot was still at the head of the pack of twenty candidates, though not always for reasons easily quantified. He clearly had the aptitude to fly the mostly automated Vostok spacecraft and was physically fit. He was short—helpful when being jammed inside the tiny capsule—and had enough experience to demonstrate the mental fortitude required for the upcoming short flight into orbit. He was also able to endure Soviet cosmonaut training, as elaborated upon above. But there was something else hinted at as people described him ... something more akin to a description that some journalist in the Capitalist West might use.

Yuri Gagarin had *star power*.

**Figure 1.1. Yuri Gagarin: cosmonaut tough, movie star handsome.**

And what exactly is star power? We all know what it looks like—Angelina Jolie, Lady Gaga, and Brad Pitt all qualify. But plenty of other beautiful movie people do not have it—so it's not just looks. Nor is it simple intellect—witness Britney Spear's infamous quip about not wanting to visit Japan because they ate too much fish, and thinking that it was located in Africa.[2] No, star power is that *thing*, that *weird glow*, that *presence* that some people just have. A presence that seems to suck some of the air out of every room they enter and replace it with a low, throbbing energy. Young Yuri Gagarin had that—and it was part of what propelled him into orbit.

Gagarin's origins would not have indicated that becoming the most famous man in Soviet history was an assured thing, however. Far from it. Like so many in the USSR, his family, already of very modest means when the Germans invaded the Soviet Union in 1941, had been decimated by World War Two.

Yuri Alekseyevich Gagarin was born in 1934 is the small village of Klushino, about one hundred miles west of Moscow. Despite its proximity to the capital city, the Gagarins were village folk, though Yuri's mother, Anna, had metropolitan roots. Anna had grown up in St. Petersburg and was a well-read woman, a passion she passed on to young Yuri, his two brothers, and his sister. Yuri later said of his mother, "I owe her everything I have achieved in life."[3] Yuri's father could not have been more different. He was a taciturn man of humble origins who was extremely practical and wanted his son to follow him into carpentry.

Before the war the family worked on a collective farm. By the end of the conflict, the German army had crisscrossed the country. The first impact on Gagarin's village was the arrival of Russian refugees heading east; then came the retreating Red army. The Germans followed in 1942, and the family's already difficult life became far bleaker. The Gagarins were forced out of their modest home and lived the remainder of the war in a small shelter dug from the earth—a ten-foot-by-ten-foot hut made of sod and scrap lumber. The Germans sent his two older siblings to a slave labor camp in Poland, where they remained there through the end of the war, returning home only after its conclusion.

Obviously, schooling for any of the children was out of the question under such circumstances, other than what basic instruction Anna could provide under the wretched conditions. Yuri, however, remained boyishly buoyant and, when the family relocated to a village called Gzhatsk in 1946, he resumed his education under the most primitive of conditions, with a volunteer teacher instructing the children using scraps of paper and charcoal left over from the near-destruction of the city.

**Figure 1.2. The Gagarin home in Gzhatsk (now renamed Gagarin), which has since been made into a museum for Gagarin. (Courtesy of Wikimedia Creative Commons, author: Kastey, licensed under CC BY-SA 3.0.)**

His father, however, was unimpressed with the idea of further education for Yuri—he wanted his son to practice a practical trade. But Yuri would not be dissuaded, and at age fifteen he left home to live with an uncle in Moscow, where he soon became an apprentice at a steel plant near the city. But while he was competent in the work, his heart was not on the ground with his co-workers, but rather in the sky. During the war, a Soviet fighter plane had crashed not far from the village, and when another plane landed to rescue the downed pilot, Yuri got his first look at the future. The pilots were impressed by Yuri's curiosity and spent time with him, even allowing him to sit in the cockpit of their plane. He never forgot the experience, and from that time on a passion for flight grew within him.

After secondary school, Yuri enrolled in a technical institute, learning, among other things, tractor repair. On the weekends, however, he was able to indulge his love for flight, training at a local flying club and learning to pilot biplanes and a single-wing trainer. After Yuri graduated from the technical institute in 1955, he was drafted into the Soviet Army and eventually sent to the First Chkalov Air Force Pilots School. Upon his graduation in 1957, he made two life decisions: first, he chose a posting with the Northern Fleet in an aviation unit near Norway, and second, he got married.

The new Mrs. Gagarin was Valentina Goryacheva. The couple met at a school dance, and, while Valentina had been seeking someone older and more sophisticated, she found Yuri to be charming, curious, and engaging. He was also self-assured, which she admired. For his part, Yuri fell for Valentina quickly. He was soon embraced by her family and he spent many happy hours reveling in their hospitality—until he took their daughter away with him to his frigid, arctic flight posting.

While settling into his new quarters—a small bachelor officer apartment he shared with his wife, as there were no units for married couples—Yuri was quite intrigued by the launch of Sputnik, the world's first satellite, which had been launched from Russian soil. Aside from his pride that the battered Soviet Union had risen from the ashes of war and accomplished this amazing feat in just over a decade, he was also impressed that spaceflight seemed to have come one giant step closer—this was, to him, a natural extension of flying. But he did not, at that time, see a place in it for him—surely men in spaceships were at least a decade away.

**Figure 1.3. Yuri Gagarin with his wife, Valentina (*left*), in 1966. (Courtesy of Wikimedia Creative Commons, user: Eivag333, licensed under the GNU Free Documentation License.)**

In 1959, in the murk of the slowly receding arctic winter, the first of the couple's two daughters, Yelena, was born. Their second, Galina, would follow in 1961. But the life of the Gagarins would be changed by much more than mere parenthood in the intervening two years.

Later that year, groups of Soviet officials appeared, unannounced, at various air bases across the country. Pilots were taken into small rooms where they were asked many questions but given no answers to their own—something special was coming, but nobody would say what it was. At least three thousand candidates were interviewed, and that number was soon reduced to a couple of hundred, and then about 150. Select candidates were sent to Moscow for medical tests and further examination, the results of which winnowed the group down to just twenty.

The tests were similar to those being endured by American test pilots across the Atlantic, and with an identical goal—to find the best physical and mental specimens for the first human voyage into space. The process was not pleasant, as anyone who has seen *The Right Stuff* (or better yet, has read Tom Wolfe's excellent book) knows. The candidates underwent batteries of psychological testing, frequent blood draws, and invasive dietary restrictions. Their eyesight was thoroughly tested—color-blindness, focus, night vision, and anything else the doctors could think of was checked. The young men were placed in sealed chambers and subjected to various atmospheric pressures (simulating high altitude) and temperatures (ditto) for increasing periods of time. After these tests, a still smaller group of finalists was selected.

Yuri and his fellow remaining candidates were finally let in on the secret, but forbidden to tell anyone, including their spouses—they were being groomed to be the Soviet Union's first cosmonauts. In this, the program diverged from the US efforts substantially, where the first astronauts were soon all over the news as the "Mercury 7."

In early 1960, the new cosmonauts were relocated along with their families to Moscow, and it was here that the most demanding part of their training would begin. While they had been subjected to arduous tests during the selection process, they would now train rigorously to become even better adapted to the expected vicissitudes of spaceflight. Physical conditioning was paramount, as was procedural training such as parachuting and flight simulation. Classes in spaceflight theory were added to the regimen, along with physical science and some medical education. It was far more preparation than they would need to complete their missions, but at the time, nobody was quite sure what would be required. As in America, the original cosmonauts were over-trained and over-conditioned. Their superiors went so far as to modify an elevator in a tall building for high-speed drops to simulate very brief periods of weightlessness and rapid slowing— a typically pragmatic (and thrifty) Russian approach to zero-g simulation. In contrast, the United States used a modified jet airliner, the "Vomit Comet," which flew at a steep angle upward, then dived for about a half minute to offer far more convincing simulations of weight-less conditions.

There was also centrifuge training to simulate the extreme gravitational forces of launch and reentry, and endless hours spent in local sports facilities for maintaining fitness. It all seemed a bit over the top to the trainees, but, with some doctors fearing that even these superb physical and psychological specimens would lose consciousness or descend into short-term insanity upon flying into space, it was considered a worthwhile effort. Nobody wanted the first spacemen to return home as gibbering lunatics—or dead. The entire enterprise was a leap into the unknown.

Figure 1.4. Gagarin in training garb about a year before his epic flight. (Courtesy of Russian News Agency TASS.)

The training was rigorous but brief—the Soviets needed to quickly select two cosmonauts, a primary and a backup pilot, for the first flight—it was imperative to beat the Americans with a man in space.

More classroom work in astronomy, rocket and spaceflight theory and technology, geophysics, and specialized medicine, followed. Much of the study was in fields that were new and, at the time, highly theoretical, and Gagarin later said that it was "a case of the blind leading the blind."[4]

Out of the twenty candidates, the two finalists were Gagarin and Gherman Titov, selected due to their performance in tests and training, as well as their physical stature. The Vostok capsule was cramped, and with Gagarin standing at five feet, two inches tall, he would fit nicely into the small ejection seat. Gagarin was also the runaway first choice by his cosmonaut comrades in a peer vote. While still only in his mid-twenties, his perseverance and tough adaptability was forged from years of hardship as a teenager and young adult, and this impressed his fellow cosmonauts.

From the Soviet leadership's perspective, it also did not hurt that Gagarin came from the humble peasant roots of collective farming—a background similar to Soviet premiere Nikita Khrushchev's. Titov had been reared in the equivalent of a middle-class environment, and in Soviet doctrine, people with "commoner" roots were preferred. Gagarin was assigned to the first flight of Vostok, with Titov as his backup and also selected for the second Vostok flight.

April 12, 1961, was Vostok 1's scheduled launch day at the Baikonur Cosmodrome Site Number 1 in southern Kazakhstan. Both Gagarin and Titov arose at 5:30 a.m. Moscow time and suited up—the politicians overseeing the flight were not taking any chances and wanted a backup pilot ready to go on a moment's notice. The pair rode in a bus to the launchpad, where the modified Soviet R-7 ICBM sat, topped by the new Vostok spacecraft, all of it held vertically by four long trusswork arms that would swing away toward the ground at the moment of liftoff. The rocket was 127 feet tall, pointing purposefully at the sky into which it would soon ascend. Gagarin thought it was a beautiful sight.

As Gagarin stood at the base of the rocket, which was already fueled and venting wispy clouds of liquid oxygen boiloff, the chief designer, Sergei Korolev, approached him. Korolev gave Gagarin a warm sendoff, telling him that he hoped to see the young man walking on the moon within a few years. Gagarin then went to the elevator, waved to the assembled technicians and officials, gave them his ten-thousand-watt smile and said, "See you soon!" and rode the elevator to the waiting Vostok capsule. He gave every appearance of a young man headed off to the market or a date with his girl, not someone attempting to create history on a very dangerous rocket.

Once inside the capsule, Gagarin's suit was hooked up to the life support system, and he went through communication checks. His call sign for the flight would be *Kedr*, or cedar, a tree which has a long history in Russian folklore. While the Vostok was intended to be fully automatic in operation, in an emergency the pilot would be able to override the controls. The American Mercury capsule was similar in this regard. But most cosmonauts—and astronauts—prized their training as pilots and favored manual controls whenever possible. It was a struggle that lasted throughout the Soviet program during the space race.

Gagarin chatted via radio with personnel at the control center while technicians prepared the spacecraft and dealt with a possible sealing problem with the hatch. Korolev, pacing in the control room, fretted over the coming launch—the R-7 booster (from which the Vostok booster was derived) had a roughly 50 percent success rate. By coincidence, that was about the same operational failure rate as the American Atlas missile that would later carry US astronaut John Glenn into orbit.

There was no dramatic countdown as there was in the American rocket program (a legacy of the Germans who designed the procedures used in the United States). Instead, when the clock reached the appointed time (and without any problems that would delay the launch), a button was pushed and the rocket departed. It was all much less fussy than their competitors half a world away, and usually just as effective.

At 9:07 a.m. Moscow time, twenty engines ignited at the a base of the rocket (augmented by a number of smaller, steerable rockets), the big swing arms arced free, and Gagarin was on his way amidst great gouts of smoke. As the rocket flew skyward, Gagarin said, "Poyekhali!" over the radio, an exuberant "Let's go!"

Two minutes after launch, the four side boosters fell free, and Vostok 1 continued in a curving trajectory toward orbit, lofted by its center engines.

**Figure 1.5. Vostok 1 lifts off on April 12, 1961, carried aloft by a modified R-7 ICBM booster.**

A minute later the launch shroud (nose cone) split and fell away from the upper stage, exposing the Vostok spacecraft and allowing Gagarin a view through a small window. The upper stage ignited, completing the powered flight and injecting Gagarin into his one-orbit flight. Gagarin radioed down, "The flight is continuing well. I can see the Earth. The visibility is good. I almost see everything. There's a certain amount of space under cumulus cloud cover. I continue the flight, everything is good."[5]

The final stage fired for about three minutes, and Gagarin said, "Zarya-1, Zarya-1, I can't hear you very well. I feel fine." Zarya-1 was the code name of the tracking station. "I'm in good spirits. I'm continuing the flight."[6]

He would soon fade from radio range. The Soviets had not yet implemented a global tracking system like the United States had designed, with dishes in various countries and at sea—for the earliest Vostok flights, all tracking stations were across the territory of Soviet Union. But the country measured over six thousand miles across, about a quarter of the Earth's circumference, so this made more sense than it might seem.

At 9:17 Moscow time, the Vostok spacecraft separated from the upper stage, and Gagarin reported that everything was proceeding as planned. He also kept up a running commentary to allay the fears of the doctors that something untoward might occur to the first human in space—"I feel splendid, very well, very well, very well."[7]

About a half hour later, Gagarin reported flying into the night side of the Earth. At 10:00 a.m. Moscow time, an announcement of the successful launch and orbital flight was made by the Soviet news agency. At home, his mother wept upon hearing the news—it was the first she knew of it, as Gagarin's mission had been kept from her and even from his own wife.

Gagarin recalls being neither hungry or thirsty during the short single-orbit flight, but it was part of the assigned tasks on his checklist, so he made sure to drink some fluids—the doctors were concerned that humans might not be able to swallow in zero-g.

Throughout the flight, a TV camera aimed at him transmitted low-resolution images back to Earth, which were being watched carefully by Soviet scientists—and the CIA. Everyone wanted to see how the first human in space was faring.

About seventy-nine minutes after launch, the Vostok's automated systems aligned the spacecraft for retrofire. Small rockets would ignite to slow the craft sufficiently to allow it to fall back into the atmosphere at the proper time to land somewhere in Soviet territory. At about 10:30 the retrorockets fired as Vostok flew over the west coast of Africa. After about forty-two seconds these engines shut down, and the reentry module—the part that Gagarin was in—separated from the power and propulsion module . . . mostly. A small bundle of wires failed to detach, and as the capsule began descending into the atmosphere, it was dragging the larger secondary module behind it. The reentry module was spinning slowly due to drag from the trailing hardware.

This was a real danger. Although the orbital module was a sphere, unlike the cone-shaped US spacecraft, it still needed to be oriented properly to reenter safely. With the power and propulsion module still connected, it could be tugged into the wrong attitude and Gagarin could be incinerated. Gagarin was not sure what was wrong, but he knew he had problems. "The wait was terrible," he later said, "It was as if time had stopped."[8]

As the drag increased, and the high temperatures caused by friction with the ever-thickening atmosphere rose, the attaching wires melted, but not before causing some wild, sickening gyrations of Gagarin's capsule. When the wires finally parted about ten minutes after reentry, Gagarin's capsule spun free, tumbling rapidly and generating about ten g's of gravitational force. He came close to blacking out.

But Gagarin managed to remain conscious and continued trying to transmit his status to the ground. At an altitude of a bit over four miles, the hatch blew free of the capsule and the ejection seat that Gagarin was strapped into rocketed away from the Vostok. While Soviet designers knew the capsule would probably survive touchdown, the Vostok was designed to return to dry land within Soviet territory, and the impact would be a hard one. With Gagarin removed from the spacecraft, and descending with his own parachute, he would land safely about two miles from the capsule. This system was revised after the Vostok flights to a more conventional—and surely less thrilling—inside-the-spacecraft landing, cushioned by braking rockets.

Gagarin landed in rural farmland about five hundred miles southeast of Moscow. His descent was witnessed by a farmer and his daughter, who ran over to see what had come from the heavens. They were shocked at the sight of a man in a bright orange spacesuit with a bulbous sphere atop his head, and Gagarin raised a hand to them, saying, "Don't be afraid. I am a Soviet like you, who has descended from space and I must find a telephone to call Moscow!"[9]

In the meantime, some local boys who had seen the capsule come down with a loud thump, cautiously approached the strange ball from space, peered into the gaping hatch, and by the time officials arrived had helped themselves to some of the uneaten space food that remained inside. Boys were the same everywhere, it seemed.

Gagarin was retrieved by military recovery forces and taken to a nearby airfield where Titov and others awaited him. He spent two days in relative calm, chatting with Titov and being evaluated by doctors, then returned to Moscow, where a public relations storm was brewing. The Kremlin announced through official news channels, "This achievement exemplifies the genius of the Soviet people and the strong force of socialism."[10] While somewhat hyperbolic, it was not far from the truth.

In Moscow, throngs of people awaited his return. Gagarin had left Earth on a secret mission and returned as an international celebrity—for a time, the most famous man on the planet. It seemed that everyone wanted to touch him, to hear him utter a few words, to share in his success. He later recalled, "I was ready for the trials of outer space, but I was not prepared to meet that sea of faces."[11]

Shortly after arriving in Moscow, Gagarin was being driven to Red Square, where his family, Premiere Khrushchev, and hundreds of thousands of people were waiting for a view of the first spaceman in history. And thus began his public relations campaign for the Soviet Union. It would last, in one form or another, for the rest of his too-short life.

Gagarin was just twenty-seven years old, and was instantaneously thrust into global fame. But, even without formal training in PR, he performed brilliantly, using his naturally sunny disposition and good looks to charm audiences everywhere. It was not an act—while he did tire of it over time, he felt that he owed it to his country to offer the best of himself at endless speeches, dinners, and official ceremonies. He remained humble and kept his sense of humor throughout. A few of the more taciturn American astronauts who would reach fame in the next couple of years probably envied him these abilities.

**Figure 1.6. Gagarin's flight made headlines across the world— including in this newspaper from Huntsville, Alabama, where America's own space program was taking shape. (Courtesy of NASA.)**

Gagarin did, however, suffer somewhat later that year. The pressure was getting to him, and, during a vacation in a Russian resort, he got drunk, wandered into the room of a nurse, and was still there—and somewhat indisposed—when his wife later came looking for him. Despite his months of training and having pulled off the world's first manned spaceflight with supreme calm, he panicked and jumped out of a second-story window. He was rather severely injured and was barely conscious when he was taken to a hospital.

Gagarin recovered, but he began drinking heavily, and his mood soured. He was still fulfilling his outreach duties but with less fervor than before. By 1963 he was in charge of training the cosmonauts for upcoming flights. But what he really wanted was not to be helping others to achieve this goal—though he performed that job well enough—but to be in training for another flight himself.

This was, unfortunately, not to be. The Soviet leadership wanted Gagarin to remain a national hero, a symbol of Communist genius and might, and assigning him to another spaceflight would be risky. It would be bad enough to lose *any* cosmonaut, but to risk losing Gagarin was unthinkable. He was the victim of his own success.

Nonetheless, by 1964, Gagarin was doing his best to prepare for the possibility of another flight assignment by studying engineering and science, subjects that many of the other cosmonauts knew much more about than he did. He managed to eventually work his way back into active cosmonaut status.

As 1966 drew to a close, the Soviet space program was about to make a major shift. The single-seat Vostok had been replaced by the larger Voskhod spacecraft, which only flew twice—first with a crew of three (to set a record for number of crew members), and later with a crew of two, one of whom would make the first spacewalk in history (setting another record for the first spacewalk). Now, as 1967 neared and with the two Voskhod flights behind them, preparations were underway to launch Russia's newest triumph, the three-seat Soyuz spacecraft that would soon carry triumphant cosmonauts to the moon.

The new spacecraft had experienced severe developmental problems, however, and the scheduled launch in April 1967 had many people deeply concerned, including the cosmonaut scheduled to fly the spacecraft, Vladimir Komarov, and his close friend Gagarin. Numerous attempts were made to tell the Soviet leadership that it was too soon; that the Soyuz was not ready for flight—it was riddled with technical problems. But the new Soviet premiere, Leonid Brezhnev, and other party hacks were unmoved. A space feat was desired for the anniversary of Lenin's birthday on April 22. The flight would take place as scheduled, despite the 230 or so faults the engineers had found with the spacecraft.

Gagarin was concerned for the life of his friend Komarov and for the program. Having gotten himself back onto active status as a cosmonaut, he began some political maneuvering to have himself assigned to the flight, saving Komarov from possible death—Gagarin knew that the Soviet leadership would never risk his own life on the flight of Soyuz 1, and that the mission would likely be delayed until the spacecraft was made safer. But Komarov refused the gesture, thanking Gagarin, and in any case the Politburo was not interested in Gagarin's concerns.

Komarov launched on April 23 and was dead within a day, killed after a trouble-filled flight that ended with his parachutes failing to open properly. This disaster stalled the development of the Soyuz, which complicated the Soviet Union's desires to beat the American Apollo lunar landing program, and also shattered Gagarin's chances for another flight assignment for good—he was officially barred from future spaceflights.

Nonetheless, Gagarin kept up his training. Whether it was to maintain the respect of the cosmonauts he oversaw, or whether it was principally to someday regain active flight status, we cannot be sure. It was probably a combination of both.

In early 1968, Gagarin completed a program at Russia's premiere aerospace training facility, the Zhukov Air and Space Defense Academy. He then decided to get requalified in modern jet fighters, which had come a long way since his earlier flight training in the late 1950s.

On March 27, 1968, Gagarin was flying with his instructor in a MiG-15 fighter as part of his flight qualification. But the weather began to worsen shortly after takeoff, and the instructor, himself a highly decorated Soviet pilot, cut the training short, heading back to the airfield after only about five minutes. But they never made it. There was a muffled boom far from the landing strip, and would-be rescuers later found the remains of the crashed MiG and the two pilots deep in the forest. A driver's license in the pocket of one of the bodies identified with certainty that it was Gagarin, and with it was a photograph of Sergei Korolev, the old chief designer who had befriended Gagarin early on, and who had died in 1966. Of this relationship between Gagarin and Korolev, a fellow cosmonaut from that early group, Boris Volynov, once said that "Korolev treated him as if he were his own son."[12]

Subsequent investigation indicated that Gagarin and his instructor were trying to pull out of a sharp dive before they crashed. Speculations over what caused the dive have ranged from another aircraft that passed too close to Gagarin's MiG, misplaced weather balloons, government sabotage, and even UFOs.[13] It is impossible to say with assurance which of these was the case (with the exception of the last), but we do know that something forced the MiG to lose altitude quickly enough that the pilots were unable to recover from the fatal dive.

Gagarin was thirty-four when he died, leaving behind Valentina and his and two daughters. His passing was deeply felt throughout the Soviet Union and internationally, and even the Mercury astronauts expressed heartfelt condolences for a fellow space voyager. Gagarin had accumulated a vast array of honors during his short life, including Hero of the Soviet Union, the Order of Lenin, the Hero of Socialist Labor, and well over thirty other major national awards.

Just over a year later, Neil Armstrong and Buzz Aldrin landed on the moon, effectively putting an end to Soviet efforts to win the space race. In 1971, the Apollo 15 astronauts carried to the moon a small statue called the Fallen Astronaut, and left it near their landing zone when they departed. The 3.3-inch-high aluminum sculpture was a memorial to the astronauts and cosmonauts who had died in their nations' attempts to conquer space and reach the moon. A small plaque left nearby listed the fourteen men who had perished in spaceflight, with Gagarin's name being sixth on the list. The plaque will remain there, barring any cataclysmic events, for billions of years.

# CHAPTER 2

# JOHN GLENN: THE CLEAN MARINE

Tipping the wing of his Vought F4U Corsair fighter, John Glenn looked back at the tiny Pacific atoll behind him where he had just dropped an incendiary bomb. The target had been a small Japanese-held town on the Jaluit Atoll in the Marshall Islands. Glenn was transfixed by the horrible beauty of the large, orange-red, expanding ball of flame that erupted on the tiny island, and horrified by the cruel, destructive effects of his deed. Napalm is a gasoline-derived jelly that sticks to anything it comes in contact with and burns furiously—a horrible way to die, he thought to himself. "Flying in combat, you don't look into the eyes of the enemy you are trying to kill," he would later write. "But napalm was a hideous weapon, and it made you think. . . . We were fighting a war we hadn't started, for the survival of our country, our families, our heritage of freedom. . . . Using napalm did not fit with peacetime sensitivities, but peace and a return to the sensitivities that it permitted were what we were fighting to achieve."[1] It was 1944, and Glenn was well into his yearlong tour of duty as a fighter pilot in the Pacific Theater of World War Two.

Glenn was a warrior with a conscience. He was also a patriot and a realist. Staunchly religious yet deeply curious, frugal yet generous, joyful yet conservative, and a person who ultimately believed in taking full responsibility for his actions. He was truly one of the good guys, and an icon among the early astronaut corps. During his long life he steadfastly navigated the rocky shoals of warfare, spaceflight, and perhaps the most daunting undertaking of them all, politics.

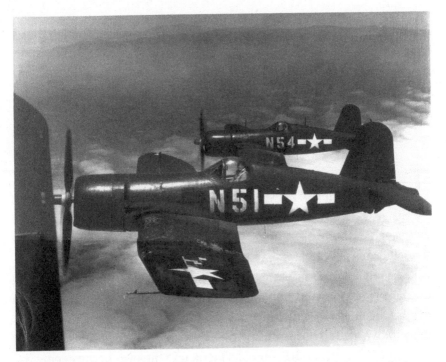

**Figure 2.1. John Glenn in a F4U Corsair fighter, near the end of WWII in the Pacific theater. (Courtesy of the United States Air Force.)**

Glenn was one of only two American astronauts to serve in combat in World War Two (the other was Deke Slayton). His postwar service would continue into the waning years of Republican China before the revolution that made the country staunchly Communist in 1949, and into the Korean conflict. He commanded squadrons and was hit by antiaircraft fire more than once. For his efforts he received a number of medals, including five Distinguished Flying Crosses. But these are not the measure of who Glenn was, just decorations for his valor. He was much more than just a "clean marine."

John Herschel Glenn Jr. was born on July 18, 1921, in Cambridge, Ohio, nestled in the northwestern extreme of the Appalachian Mountains. His father was a veteran of World War One who married shortly after returning from that conflict. John Glenn Sr. had little formal education, departing school after the sixth grade. He worked at a number of jobs before becoming a plumber and moving his young family to New Concord, Ohio, where he opened a plumbing supply company. Glenn Sr. would continue turning a wrench while his wife operated the store, and this became the backdrop for John Glenn Jr.'s upbringing.

New Concord was a model of small-town America for the era, Glenn would later say, a town where "Boys learned the company of men—the way they talked and held themselves, and their concerns—at the town barbershop and hunting in the woods."[2] Glenn recalled his childhood as almost idyllic, in a way that could only happen in small town America.

By the age of eight, Glenn was already experienced in small town and country life, working hard in the family business and odd jobs, but he had not yet had a taste of life in the air. That changed one day when he accompanied his father on a plumbing job. As they drove back from the job site, his father spotted an airplane parked in a grassy field outside of town—still a rare sight in those days. He decided to stop and indulge both his own and his son's curiosity.

The craft was an old open-cockpit WACO biplane of a design that was popular as a mail-carrier and for other civilian services. The pilot was "barnstorming"—offering short rides in exchange for payment—and reminded Glenn of the dashing pilot-adventurers he saw in the Sunday comics in the newspaper. Glenn Sr. wanted a ride and told his son, "If you don't want to do it, I'm going anyway. So you better come unless you want to sit down here and watch."[3] That was all the prompting the younger Glenn needed, and he rode in his father's lap in the second cockpit. The experience transformed him. Not only had there been the adventure of doing something new and daring, the view from high in the air, and the thrill of the wind buffeting your face in the open cockpit, but there was also the *newness* of it—a glimpse of the future. Glenn felt something he later realized that he shared with his father—"His eagerness to experiment was one of the most important lessons of my youth," Glenn would later recall.[4]

Soon, his bedroom was filled with model airplanes—not the plastic kind that dominated young people's rooms in the 1950s and beyond, but wood-and-tissue paper miniatures that were powered by a rubber-band-driven propeller. He built and flew dozens—he was, after all, growing up in Ohio, the home of the Wright Brothers, a fact he was keenly aware of throughout his youth.

Glenn was desperate to learn how to fly, but the times were not right for such undertakings. The largest financial disaster in American history, the Great Depression, was crippling the nation, and his family was feeling the pinch along with most others. Frivolous ideas such as flight instruction went out the window as Glenn pitched in to help make enough money to keep a roof over their heads. The family leased plots of land to grow their own food, and Glenn spent much of his off-hours tilling a small plot of vegetables.

Frustrated by the limitations of the era, Glenn, along with a pack of other young boys, formed a group they called the Ohio Rangers, modeled after the Boy Scouts, which did not have a presence in New Concord. They would rove beyond the town's borders and into the forests, first to hike and explore and later to hunt. They built their own camp out in the woods where they would spend weekends and summer days and nights, camping, fishing, and hiking, and learned to be independent and self-reliant. Their camp had spots for tents, and they went so far as to carry in loads of sawdust to create a "parade ground," a standard, military fixture of the era. There, in their rude replica of a 1920s US Army base, they could stand in formation and pay homage to the American flag, which they hoisted up onto a tall sapling. Glenn even brought along a bugle to play taps as they brought the flag down in the evenings. This was a true slice of Norman Rockwell's America.

Two other huge influences were also shaping his life at the time. The first was the state fairs that his father never tired of taking him to, where Glenn was fascinated by the science exhibits put on by various schools. He endured touring the hog pens and prize cow exhibits so that he could spend time looking at science experiments. The highlight of these visits came at the 1933 Chicago World's Fair. World's Fairs were enormous affairs, with huge buildings and plazas built just for the annual event in cities all over the world. True to form, Glenn's father bundled his son and three adult friends into the family's tired Chevy convertible and headed off to Chicago to see the newest in science, technology, and aviation. The event did not disappoint, and it left a lasting impression on Glenn—the future was bright, and much of it would take place in the air.

The second influence was the love of his life, Annie Castor. Like something out of a fairytale, he met young Annie when they were both toddlers, and they were fast friends throughout their childhood. Annie was cute, athletic, musically gifted, and utterly charming in Glenn's eyes. She also had a severe stutter, something that would over time create huge social anxiety for her—she was so thwarted in her attempts to communicate that speech would often fail her altogether, and she resorted to carrying a notepad and pencil to afford communication with others. But Glenn was not bothered by this, and was smitten at an early age, to the extent that, according to him, he never looked seriously at any other girl.[5]

By the time Glenn was midway through high school, he had developed a strong interest in chemistry. He began spending time with a young man from the local college named Julian White. Julian was a chemistry major, and Glenn would join him at the college lab on the weekends, working with Bunsen burners to heat and form glass for experiments, and solving simple chemistry problems. It was that Glenn family curiosity at work, something that would serve him well in the future. Glenn also had a physics teacher in high school who engaged with him in ways that stimulated his interest.

Glenn went out for the football team during his freshman year, but was not as successful as he thought he might be. Being small and moderate of stature, he was destined to play center, a position that assured a solid trouncing far too often for his tastes. He moved to the basketball team, and eventually added tennis to his repertoire, as well as glee club. That last inclusion was so that he could spend time with Annie at school, and he never regretted the decision, despite occasional ribbing from the other boys.

But the Depression was still grinding on, and when he was not playing sports, Glenn spent much of his free time working with his father on plumbing projects—digging sewer line trenches and septic toilet pits. It was not work he enjoyed, but he did it alongside the other adult workers his father had hired and held his own.

The only time Glenn recalls getting really angry in this period was when he heard that a number of girls at the school were making fun of Annie and her speech impediment. "Annie's stuttering wasn't something I viewed as a problem," he would recall. "It was just something she did, no different from some people writing left-handed and others right-handed. I thought it was cruel and thoughtless to laugh at someone for something like that . . . and I told them so."[6] The mocking ceased instantly.

After graduation, Glenn knew he wanted to go to college, and despite the family's limited resources, his parents supported the notion. Glenn was already feeling the pull of the road—he wanted to broaden his horizons—but an out-of-town college would cost much more than a local one, so he decided to apply to nearby Muskingum College (now Muskingum University). It was not his first choice, but between the savings in room and board by living at home, and receiving a scholarship, it was the best option.

Having enjoyed his weekend work at the college during high school, Glenn decided to major in chemistry, thinking that he might become a researcher or go to medical school. An added advantage was that Annie was already in attendance there, one class above him. He enjoyed his studies, but flying still gripped his imagination and simply would not let go—he needed to find his way into the air.

In his sophomore year, the war in Europe was heating up—Glenn was moved by the impassioned speeches of Winston Churchill as Britain struggled to repel the German invaders coming across the English Channel in waves of bomber aircraft. In the first days of a new term, right after Christmas break, Glenn saw a notice posted at school. It was from the Civilian Pilot Training Program, and Glenn could barely believe what he was reading. The program would pay for the cost of ground training and air instruction for those who qualified. To a kid raised during the Depression, this seemed almost too good to be true. The program also awarded college credit for specific aligned subjects, and upon completion you were granted a private pilot's license.

Glenn ran to the physics department, where the program was being hosted on campus, to join the program. There he discovered one additional catch—by signing up, you were assuring the government that you would be available for military flight training "when needed." That was not a problem—a draft could be looming anyway, given how world affairs were evolving—so he signed up and was accepted within weeks.

Within a few months Glenn began the commute to the nearby airfield where the flight training would occur using a small single-engine plane called a Taylorcraft, which had all of 65 horsepower. But it was an *airplane*, and that was all that mattered.

Glenn excelled in the air, with an almost obsessive attention to detail and careful awareness to everything going on around him—both excellent attributes for pilots. Just weeks later he was flying solo, and by July he had his license.

That summer was all about finding ways to try and keep his hand in at the airfield. By the time his junior year loomed, however, the bad news out of Europe was casting a pall over the country. Hitler's *blitzkrieg* was devouring one nation after another, and the United States was becoming increasingly anxious.

A few months later, in December 1941, fate intervened to change his life forever.

Glenn was on his way to a concert in which Annie would be performing on the organ when he heard the news of the Japanese attack on Pearl Harbor. He greeted her before the concert, and tried to enjoy her performance, but by the time it was over he could no longer hold in the news. America had suffered a sneak attack, thousands were dead or dying, and war had been declared. He knew what he had to do, and so did she.

Soon thereafter he signed up for the Army Air Corps, the precursor to the postwar air force. When an order to appear for duty didn't come in a few months, he went to the navy recruiter instead and was mustered within two weeks. He was still in his junior year at college, and would not finish his degree there—a fact that would almost kill his career as an astronaut before it began.

Training took place in Iowa City, about six hundred miles from New Concord. Glenn took every opportunity he had to see Annie, but it was a busy time and the two were separated for longer periods than ever before. Three months later he was sent to Corpus Christi, Texas, for advanced training, and the separations became almost intolerable— he wrote her letters every day.

Not long after his arrival in Texas, the itch came to swap services. While he had joined the navy because he heard that naval aviation was more challenging (and because the army had apparently overlooked him), the Marine Corps beckoned. Glenn had heard of the exploits of marine pilots in the island battles of the Pacific, and felt that *that* kind of flying was for him—low-level air support for your fellow marines slugging it out on the ground. He would soon be headed to the Pacific theater, but there was one more task to attend to in the United States— the girl he'd left behind.

As a newly commissioned second lieutenant, Glenn hopped onto a train to return home and marry Annie. She was still wearing a $125 ring he had purchased for her before leaving Ohio. (She never did let him replace it with a larger one.) The wedding was simple and quick, and after a very brief honeymoon the new couple was off to a short stop in North Carolina, then on to Southern California.

Once in San Diego, and later El Centro, an adjacent dusty cow town with a Marine Corps Air Station, Glenn spent time in multiengine planes, then transitioned to single-engine fighters, and life got more exciting. Then, in early 1944, it was time to ship out.

He was sent to Hawaii for more training, and was then assigned to Midway Atoll, a tiny island in the mid-Pacific tenuously held by the United States. After a few months, he was sent to the Central Pacific, where he was stationed on Majuro Atoll in the Marshall Islands, only recently wrested from the Japanese. It was the summer of 1944.

The flying was always exciting, nerve-wracking at times, and occasionally very dangerous. All combat aviation is risky, but flying close air support as opposed to, say, high-altitude bombing, was some of the most dangerous. Glenn's first assignment was to fly "flak suppression." The idea was to fly in to an enemy-held island low, firing the wing-mounted machine guns as you went along to keep their heads down. This allowed the bombers or dive bombers above you to do their work relatively unscathed—you had not only "suppressed" the antiaircraft fire, but often had been one of the first targets of the enemy as well. Glenn lost his first friend to this type of combat on his first day in the war—his wingman was shot down and never seen again.

Glenn spent a year in the Marshall Islands, flying in support roles and direct attacks on nearby islands where the Japanese were fighting to save their empire. It was during this time that he had his first brush with death: while dropping bombs at low altitude over Nauru, a Japanese-held atoll, an antiaircraft shell punched through a wing of his Corsair fighter. He was not injured, and the plane made it back to the base, but he became acutely aware of his mortality in that attack.

Toward the end of his year-long duty in the Pacific, Glenn was introduced to napalm bombing. In their first experience with it, his squadron incinerated a small Japanese-held town on Jaluit Atoll. "A napalm attack had a horrible beauty ... one of the most eerie, awesome and sobering I had ever seen," he would later say. "We routinely used napalm in areas where intelligence thought there were a lot of people. It was terrible to think what it was like on the ground in the middle of those flames."[7]

After his tour he returned to the United States, and was stationed at the Patuxent River Naval Air Test Center, where as a seasoned combat veteran he would wring out new airplane designs. Glenn and his comrades flew the new fighters flat-out for hours, ironing out any bugs they could. He remained on this duty—with a brief transfer to another Corsair squadron that included the possibility of being shipped out to support the invasion of Japan—until the atomic attacks on Hiroshima and Nagasaki brought an end to the war.

Glenn had seen the world and sampled the excitement and horror of war—somehow, New Concord was just too small for him to return to. He considered various ways of making a living, but in the end, after consulting with Annie, he decided to stay in the Marine Corps.

After the birth of his first child—a son—in 1946, Glenn was sent to China. The postwar partnership between the US-supported Republican China under Chiang Kai-Shek and the Communists under Mao Zedong had unraveled instantly after the defeat of Japan. Glenn and his squadron would be in China for just a few months, it was thought, backing up the negotiations between the two sides by staging a show of strength for the Nationalists. It didn't work, and the Communists took control of the country in 1949—an event that left a lasting impression on him.

Glenn was briefly reunited with his family in Guam—he and Annie now had a second child, a daughter—then headed back to the United States. A variety of assignments awaited him there, none particularly interesting, but then war broke out in Korea. That war, for those in the air, would be fought in jets.

This represented a whole new world in aerial combat. Instead of flying at what used to seem a brisk 300 miles per hour, the new jet fighters were traveling at 600 mph. Glenn was a flight instructor in the United States at this time, watching the developments unfurl across the Pacific carefully. In February 1953, he was sent into combat again.

While in Korea, Glenn continued to write to Annie, and now to his children as well. One example, included in his memoir, was sent on the occasion of his daughter's sixth birthday, and the letter is typical John Glenn. After talking about what he had seen in Korea—all the injustice, the poverty, and the depredations he attributed to Communism, he expanded the conversation to encompass the Soviet Union and their plans for what he saw as global domination:

> Our flag represents all the things we believe our country should be. It means that children can go to school, the mothers and fathers can have homes and raise their children the way they want, and live the lives they choose. So when you see our flag, think of that.[8]

He closed with a statement about "the Communists": "Don't you worry about them ever getting to our country, because I, and many other men, are out here to see that they never do."[9] It would seem corny if he didn't mean it from the bottom of his heart.

**Figure 2.2. John Glenn's F-86 fighter in Korea. On the side is painted "MiG Mad Marine," indicating his successes in aerial combat. (Courtesy of the United States Air Force.)**

While in Korea, Glenn was hit with antiaircraft fire and shrapnel so many times that his squadron mates took to calling him "Magnet Ass" due to the quantity of dangerous enemy explosive fragments that he seemed to be attracting to his plane. Nevertheless, he escaped the conflict uninjured. Once he was rotated back to the United States, he was assigned to Patuxent River once again for more work in flight test, and was then sent to the Bureau of Aeronautics in Washington, DC, and attended classes, in engineering and aeronautics, at the University of Maryland.

Glenn had one more high-stakes goal to achieve before taking the next step in his flight career. he wanted to break the transcontinental record for supersonic flight, which at that time was three hours and forty-five minutes. Glenn thought that the Vought F8U Crusader, a high-performance jet that had premiered in 1955, could do it faster. He called this exploit "Project Bullet" because the jet's airspeed was faster than a .45 caliber bullet, the standard sidearm caliber in the armed forces at the time. On July 16, 1957, he beat the old record by about twenty minutes and was feted by the press, earning him an appearance on the popular TV quiz show, *Name That Tune*. He was the very archetype of the ramrod-straight, clean-shaven Marine Corps war hero.

Then Sputnik happened.

The United States had already commissioned an orbital satellite of its own called Vanguard, designed and managed by the navy, which was set to fly in 1958 as part of the International Geophysical Year, a global initiative in earth science that included sixty-seven countries. But the Soviets usurped US plans by lofting Sputnik in October 1957. This was a massive shock to the West, and America in particular, which had long assumed an innate technological superiority over their counterparts in the USSR. To further exacerbate the situation, Nikita Khrushchev, the Soviet leader, taunted, "The United States now sleeps under a Soviet moon." Americans didn't like that, and to Glenn it was an almost personal affront. "The American-made television sets, transistor radios, and cars with tail fins that meant technology to US consumers in the rich post-war years seemed frivolous next to the evidence of Soviet scientific achievement beeping overhead," he later said.[10]

By early 1958, Glenn was involved in research focused on manned spaceflight, the next major milestone after the lofting of orbiting satellites. Thanks to his vast experience in flying, he was invited to work with NASA's predecessor, the National Advisory Committee on Aeronautics (NACA) on orbital trajectories for manned spacecraft. Specifically, the NACA was using new digital computers—huge and underpowered by modern standards—to simulate reentry procedures.

The NACA wanted input from pilots with solid qualifications, and they got it. Glenn was tasked to sit at a simple console and model reentry trajectories. "They had a control stick, a little control toggle switch, and things like that. ... It was supposed to be [a] computer study of what orbital tracks you could get to and could not get to, and what kind of a trajectory you could make during landings." He completed the project and prepared a report for the Bureau of Aeronautics in Washington, DC. This was his first taste of spaceflight, although a simulated one. "There had been some rumors at that time that perhaps we were going to get into a manned space program some time. ... I was very interested in that."

This soon evolved into human testing for the rigors of flying in space. Glenn was sent to the Naval Air Development Center in Pennsylvania to spend some time in their centrifuge—a giant machine that simulated varying degrees of extreme gravitational forces. The centrifuge used a gondola suspended on the end of a fifty-foot arm that spun around, generating crushing forces at the end where the pilot was strapped in.

As a pilot, Glenn had already experienced four-to-six g's of force when pulling out of a steep bombing run or making a hard turn in a jet fighter, but the navy's new centrifuge could generate a bone-crushing twenty-five g's, a limit they did not expose human subjects to. Glenn was spun in up to eight or nine g's in various positions to test his ability to operate simulated controls at high g-forces—this was about the maximum that was expected during launch and reentry of a future spacecraft.

Glenn was then sent to the McDonnell Aircraft plant in St. Louis, Missouri, where he acted as the navy's representative to the company's burgeoning efforts at designing the first American manned spacecraft. This was the first of the blunt-body capsules that would be used through the end of the Apollo era—a cone-shaped spacecraft that would ride into space on a rocket, then reenter the atmosphere blunt-end first. It was the beginning of the US manned space program, though Glenn was not told so at the time. As he put it, "It consisted mostly of research and rumors."[11]

In mid-1958, the National Aeronautics and Space Administration—NASA—was created by presidential decree, and the manned spaceflight effort was now official. There had been competing efforts by the air force and navy to wrest control of such a program, but with the stroke of a pen, President Eisenhower had created a civilian space agency to undertake the program. Subject closed.

Glenn was by now thirty-seven, and still without a college degree. Between earlier college credits, the additional coursework work he had pursued at various other colleges, and his flight experience, he reckoned that he had about the equivalent of a master's degree, but no official pedigree to show for it—and he was certain that this would be required for involvement in America's new space endeavor. Nonetheless, he was determined to be a part of it by hook or by crook.

Anticipating this new call to service, and the opportunities it would afford him, Glenn looked into what the requirements would be. Due to his experience with the engineers at McDonnel Aircraft, he knew that the capsule would be cramped and tiny, and the pilots would need to be small as well. He could not do anything about his height, which was five-foot, eleven inches, but he could reduce his mass. He'd been working a desk job for some time now, and the lack of vigorous activity, grabbing meals when he could, and constant snacking had resulted in notable weight gain. He weighed 208 pounds and was determined to slim down and tone up. Glenn began a self-imposed exercise regimen both at the gym and outside—he ran, swam, jumped rope, and worked out with weights to get down to a target weight of 178 or less. He also reverted to a spartan diet to reduce his calorie intake.

At this time, NASA was still trying to ascertain what kind of people would be the best candidates for astronauts. The agency considered people with varied skill sets, ranging from gymnasts to high-wire walkers to other daredevils—anyone who might do well under extreme circumstances. After much deliberation, by the end of 1958 they had quietly decided that test pilots, preferably with combat experience, would be the best candidates.

Paralleling similar (but unknown outside the USSR) activities in the Soviet Union, NASA screened a roster of 508 test pilots and reduced it to 110 candidates. In December, NASA made the official announcement: America was embarking on a program of human spaceflight, to be called Mercury, and they would need the best of the best to fly it.

**Figure 2.3. Glenn running on a Florida beach to keep trim. (Courtesy of NASA.)**

In the first months of 1959, Glenn received a packet of orders from the navy—somewhat enticingly marked "TOP SECRET"—that instructed him to attend a briefing at the Pentagon. There, in the company of NASA officials and a clutch of other test pilots, he learned of the official plans to recruit top performers to become *astronauts*—a term none of the pilots had heard before.

Glenn volunteered immediately. "I thought this would be fascinating to do. As I saw it, it was sort of a follow on to what I had been doing at Patuxent"—his time as a test pilot. "I had been involved with our highest performance aircraft, and they were looking for people who had a lot of test time ... [I] had worked in small cockpits at high speeds. ... Another factor was that I had come back from Korea, had combat time in Korea. That was another plus." But this was not a decision he made alone. "I didn't just jump at it. I talked to Annie about it first because it was going to mean ... quite a change in just our family activity. I was going to be away from home more ..."[12]

And there was another motivation: God and country. "[This] was something I thought had a really big purpose for the country, and if I could contribute in that area, what better use could I make of whatever flight talents I might have ..."[13] He elaborated on his beliefs in his autobiography years later: "I believe strongly that each of us has a unique set of God-given characteristics, talents, and abilities. Our end of the partnership is to use those capabilities to the maximum and for a good purpose as we pass through this existence."[14] He felt that the space program would harness some of his most profound talents and abilities, if he worked hard enough at it.

He later relayed these thoughts to his wife: "Annie, the scientists I've been talking to say we have the know-how to do this. And if we can do it, we ought to. We ought to get up there before the Soviets do, if we can. ... The world has its eye on us, and if we're going to send astronauts into space, I want to be one of them."[15]

She answered, simply, "If you want to make it, you'll make it."

As the pool of astronaut candidates was being narrowed, Glenn's greatest fear loomed—his lack of a college degree had become a sticking point. However, unbeknownst to Glenn, his former commanding officer at Patuxent River had quietly gone to the astronaut selection committee to argue for Glenn's inclusion, carrying along Glenn's academic records and the details of his test flight experience in the military. He must have been persuasive—the list had by now been narrowed to thirty-two men, and Glenn would be among them.

Within weeks a letter arrived congratulating Glenn on his candidacy to advance to the next level of selection and telling him to report to an aerospace medical facility in New Mexico for further evaluation. The only other details were instructions to travel quietly, in civilian clothing, and to speak to no one of the program or his reasons for going there. He and the other candidates were given identification numbers to keep things hush-hush.

The facility was the Lovelace Clinic in Albuquerque, a place that would become infamous in a couple of decades thanks to Tom Wolfe's bestselling book *The Right Stuff*. The medical tests at Lovelace were the most extensive ever conducted . . . and included some of the most invasive procedures since the Inquisition.

Glenn related his experiences at Lovelace in a NASA oral history: "Randy Lovelace had been an air force flight surgeon. He had done a lot of work in selection processes. He had sort of this clinic out there, and it was the most advanced place to study physical characteristics of people for flight. . . . So, we went out there and they put us through all the tests, every test known, whatever they could do to the human body."[16]

> You'd come and take your shoes off and put your feet in a bucket of ice water. There had been a big study by somebody at that time, that your reaction to that, your blood pressure and your pulse reaction to that, there was a corollary with whether you were liable to develop heart problems later in life. . . . And the balance test, putting you up on chairs and spinning you around . . . vision, every kind of test they knew how to run on the human body, they did.[17]

The test pilots were a hardy bunch. But when it came to the psychological tests, for which they were moved to another facility, some found the process more trying:

> We then went to Wright Patterson Air Force Base, and there was a whole different set of tests out there. That's where we went through all the psychological tests. . . . They had an isolation test that I've never been in before or since. It was a room that was called an echoic chamber, and . . . they'd put you in there and you just sat at a desk and they turned out the lights. It was absolutely dark and it was sound proof. In fact, the room was designed so there was no sound inside whatsoever. That was the echoic part of this thing. And there you

were in the dark and completely isolated; no sound and no light and they wanted to see how you reacted in there. They had [sensor] leads on you so that they could see what your response was. They wouldn't tell you how long you were going to be in there.[18]

Glenn sat at a desk in the chamber, and after the lights were turned out, felt around for a pad of paper to occupy his time in the stygian darkness. Without being able to see what he was doing, he wrote a poem, which he later gave to Annie. She still has it.

Then came the psychological evaluations. Rorschach tests were a big part of it (we will hear later how another astronaut hopeful disqualified himself at this step), and a psychological profiling test followed, which consisted of 560-plus questions on seemingly innocuous topics asked over and over again. Some example questions:

- I like mechanic's magazines. T/F.
- I like to read newspaper articles on crime. T/F.
- I am very seldom troubled by constipation. T/F.
- No one seems to understand me. T/F.
- If people had not had it in for me I would have been much more successful. T/F.
- My soul sometimes leaves my body. T/F.
- I see things or people or animals around me that others do not see. T/F.[19]

(One suspects that the last of these questions might have caused some consternation on the part of the evaluating psychiatrists if answered "true.")

There were others that were fill-ins, with at least twenty of them starting with "I am ___" (fill in the blank). Glenn focused on answers like "I am a man." "I am a flyer." "I am a father." One can only imagine what some of the more skeptical test-takers filled in for some of these questions.

Glenn took these procedures and tests in stride; many of the other aspiring astronauts took exception to this and other parts of the process. Doctors and test pilots—or any pilots for that matter—are natural enemies; from a pilot's perspective, the best you can hope for from a doctor is to not be grounded, and psychiatrists were even more suspect.

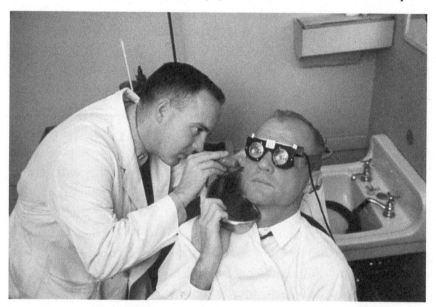

**Figure 2.4. One of the many medical tests that the early astronauts endured evaluated their sense of balance after having ice water injected into their ear. (Courtesy of NASA.)**

Two weeks after all the testing was concluded, Glenn got the call: "Major Glenn, you've been through all the tests. Are you still interested in the program?"

Was he? Of course! "Yes I am, very much."[20]

The seven selectees—who would soon be known as the Mercury astronauts, or the "Mercury 7," assembled at Langley Air Force Base in early April 1959. Three were from the air force, three from the navy, and Glenn was the sole representative of the Marine Corps. They were Al Shepard, Wally Schirra, Gordon Cooper, Deke Slayton, Gus Grissom, Scott Carpenter, and Glenn. It was a cast of alpha males if ever there was one—confident, bordering on arrogant; competent, accomplished, and very, very physically fit. Slayton was the only other World War Two veteran, but many of the others had flown in Korea.

About two weeks later, on April 19, NASA held its introductory press conference to introduce the Mercury astronauts to the world. It was by far the largest and most complex such event that NASA had held in its brief existence, and it was overwhelming not just for the management but for the astronauts themselves. "No one could have been prepared completely for the amount of attention that we got out of this thing. It felt like a tidal wave came over you," Glenn recalled.[21] "I don't think I was as tight lipped as some of the other people were, good or bad," he added.

In fact, the others started as a stiff-backed, tight-lipped group, and Glenn quickly became the media favorite due to his ability with words, his ready smile, his enthusiasm, and his downright mom-and-apple-pie-wave-the-flag patriotism. Early on, when asked how their families felt about their selection as astronauts, as the others fidgeted, a stiff Carpenter said, simply, "They're all as enthusiastic about the program as I am."[22] Next came Cooper, equally flat and monotone: "Mine are very enthusiastic also, I can answer the same for myself."

Then it was Glenn's turn. With his trademark Boy Scout smile and seeming eagerness to talk to the press, Glenn piped up, "I don't think any of us could really go on with something like this if we didn't have pretty good backing at home, really. My wife's attitude toward this has been the same as it has been all along through all my flying. If it is what I want to do she is behind it, and the kids are too, a hundred percent."[23]

The remainder perked up, possibly challenged by Glenn's performance—this was a competitive group, above all—and things smoothed out a bit. One by one the others loosened up, with Gus Grissom, sitting to the left of Glenn, and Scott Carpenter, to his right, getting some chuckles out of the press corps.

After some more answers from the other six, Glenn said, in his chipper and slightly pious way, "I think we are very fortunate that we have been blessed with the talents that have been picked for something like this..."—the other six hard cases either stared ahead or at Glenn—"Every one of us would feel guilty, I think, if we didn't make the fullest use of our talents in volunteering for something like this—that is as important as this is to our country and to the world in general right now."[24]

The press conference was just the beginning of a rapidly growing press maelstrom of interest. None of these astronauts had yet flown any higher than you could in a jet fighter, but were lauded as if spaceflight was a *fait accompli*. As the time ticked down toward that first launch, media interest was on fire. It became so overwhelming that NASA arranged for a well-known Washington lawyer with some media experience, Leo DeOrsey, to represent the astronauts as a group to the media. DeOrsey agreed on the condition that he would be uncompensated—it was simply a privilege to work with these men in this urgent American undertaking. Eventually, with the press wanting access to all seven men and their families, it was decided to make an exclusive deal with *Life* magazine, the preeminent weekly photo magazine of the era. *Life* paid the group of seven $500,000 for their stories over three years—a small fortune at the time—and while it rankled the rest of the media, that was the deal. In the end, while it may have frustrated the other outlets to get much of their information second hand, the public's curiosity was sated by *Life*'s extensive coverage of the astronauts. And the money that each family received from the *Life* deal was way more than their annual military pay.

Despite the extra money, Glenn remained a frugal family man. While the other astronauts could be flamboyant, he remained the steady hand. One example were the cars they drove. Shepard had a Corvette, at the time the fastest car in America. Schirra drove an Austin-Healey 3000, which was, after Jaguar, the sportiest British import available. Glenn, for his part, bought a virtually unknown German import called the NSU Prinz. It was tiny, noisy, dangerous, and grossly underpowered, with a two-cylinder, 35-horsepower engine (the Austin-Healey had 150 hp, the Corvette almost 300). You could drive the Prinz on a freeway, but only if you were not terribly concerned about living long enough to reach your destination. Glenn took some ribbing about the car, but he shrugged it off with a white-toothed grin. He didn't need a flashy car to announce who he was to the world.

Soon the astronauts were back at Wright Patterson Air Force Base. Centrifuge runs were considered central to their conditioning to prepare them for the strong g-forces that would be generated during various phases of the flight. Into the small metal pod they went, to be whirled at increasing speeds, resulting in ever-higher levels of stresses and disorientation. Moving slowly up through eight times the force of normal gravity, each astronaut would be instructed to attempt to operate a switch panel in front of them, only to find that their arms now felt like lead blocks. While all had experienced high-g loads in fighter jets, this was different—more prolonged, more challenging, and perhaps worst of all, there were *doctors* watching. This was not a cramped fighter cockpit, where you succeeded or failed with only God as witness; there was no privacy in their training here. The doctors quietly made their notes.

Then, to test the astronauts' reactions to the possible extra-high g-forces generated during an abort, when small but powerful rockets could be fired to whisk the Mercury capsule away from a malfunctioning rocket booster, Glenn and Cooper were subjected to more centrifuge runs at sixteen g's. At that load, a 170-pound astronaut effectively weighed 2,800 pounds. Their backs were crisscrossed with bruises and broken blood vessels for weeks afterward.

In December 1959, NASA started launching monkeys, then later chimpanzees, inside of Mercury capsules to see how primates, man's nearest cousins, would react to spaceflight. Some of the doctors were worried that humans could simply not function in weightlessness—that they would be unable to swallow or breathe, or that their hearts might stop pumping blood effectively. In flight after flight, up went the primates and down they came, none the worse for wear except for frayed nerves.

The astronauts were sent next to NASA's Lewis Research Center in Cleveland to ride one of the most diabolical machines ever designed, the MASTIF, short for Multi-Axis Space Training Inertial Facility. The machine looked like something out of *Stargate*, with a single seat inside a series of ever larger rings. Each ring could swing the occupant of that chair on a different axis. The idea was to train the astronauts to operate a tumbling capsule—something that would later occur during an emergency in the Gemini program—but this devilish contraption took it to extremes.

Into the cockpit the astronaut would go, straps restraining him tightly. The operator would spin the victim up along one axis, perhaps turning the astronaut in a forward rotation. The astronaut had a hand controller with which he would attempt to control the tumble. As he got the hang of countering the motions—each one was a test pilot, after all—the operator would add another axis of spin, until the thing was spinning in three different directions at once, completely unpredictably, and lurching violently from one axis to another. By the time they were done, the operator had them spinning at about thirty revolutions per minute.

A cot was thoughtfully provided to allow them to lie down for about a half hour afterward, or until their heads stopped spinning. A vomit bucket was situated nearby, and was frequently used.

As the physical training progressed, so did the classroom work. Astronomy was essential to better understand celestial navigation, and the astronauts practiced at a planetarium in North Carolina. As Glenn later recalled,

> They actually had set up a mock-up of the spacecraft in there. Then we had the star patterns in this planetarium the same as they would be during our flight, on the orbital flight. And so you'd look out the little window and you'd have the same star pattern. The idea of this was that, if you lost radio communications completely and you're up there, how would you get back down? How would you know exactly when to fire [your retrorockets]? Well, we had this thing, we worked on that simulator enough that, if we had lost communication and we saw the star patterns going by, you knew approximately what star patterns would be in your window. You knew where to fire to come down at a certain area, which would be at least 3,000 miles away.[25]

With the ability to control their own reentry in case the primitive onboard guidance system failed, the astronauts would also need to know how to survive in the wild in case they managed to come down in the boondocks somewhere. That necessitated survival training. All of them had endured some form of this in the military, but NASA took it to a new level—this was not combat survival training for a region of specific interest; any of them could come down in a variety of locations from hot to cold, flat desert to rugged mountains.

"They wanted us to train in survival techniques wherever we might come down, in case it was an emergency re-entry and we had to come down someplace [unexpected]," Glenn recalled. "So we had desert training. We had sea training. We had jungle survival training. ... We trained on how you live out in the jungle and how you survive for 72 hours. They said they could pick us up anywhere in the world in 72 hours. That would be the max."[26]

But Glenn was not content with just enduring the standard torture of desert survival. His ever inquisitive mind had gotten curious about the effects of dehydration on the human body. He spoke with one of the doctors supervising the training, and volunteered to conduct an experiment that most of us would avoid like the plague:

> I asked Bill Douglas, our flight surgeon . . . would he go along with it if I intentionally dehydrated and didn't have any water for the first 24 hours I was out there. I just wanted to see what it was like a little bit. And Bill said, "Yes," and he'd come by a couple extra times to make sure I wasn't getting myself in trouble. And so I did that. I didn't have any water for the first 24 hours I was out there. It was amazing to me how fast your body goes downhill in that high heat if you're not having any water.[27]

Glenn held out for a day and a night in the desert heat.

> At the end of that time period, Bill wanted me to drink as much water as I wanted, and if I remember the figures correctly, for the next nine hours I drank 15 pints of water and didn't have any inclination to pass any on at all. So you can really get yourself in bad shape in a big hurry. That really impressed me.[28]

As the astronauts prepared themselves to fly, NASA's workforce pre-pared their machines. The monkeys had flown, but now that humans would be going up, safety was far more critical. The Mercury capsule would be flying on two separate rockets. The first two manned flights would be lofted by the Redstone booster, a small intermediate range nuclear missile that had been deployed by the army in Europe a few years earlier and had proven to be quite reliable in tests. The problem was that the Red-stone only created about 70,000 pounds of thrust—somewhat less than twice what a single 747 jet engine creates—and would not be capable of lifting the Mercury capsule, which weighed just under 3,000 pounds, into orbit. That duty would be fulfilled by the larger Atlas booster, another repurposed nuclear missile.

The Atlas was an entirely different animal. While the Redstone had been derived from the German V2 missile of World War Two infamy, the Atlas had been created from scratch by US designers. It was an odd beast, made from exceedingly thin stainless steel—essentially just a nose cone on top of fragile metal tanks. The fuselage was so thin that the rocket could not stand under its own weight unless pressurized; it could crumple like a stack of wet paper bags if not handled properly. While it was an effective early nuclear warhead carrier, the fragile rocket did not inspire a lot of confidence for carrying astronauts into orbit. But the United States knew that the Russians were forging ahead to put their cosmonauts into space with their larger, sturdier rockets, and the Atlas was all that was available in the American inventory. It would have to do.

Early in their training, the Mercury astronauts had been invited to see a test launch of their chariot to the stars. They arrayed themselves in the viewing area at the launchpad at Cape Canaveral, expecting to witness the glory of a rocket launch, something none had seen before. At the appointed time, the base of the Atlas rocket was engulfed in orange flame as its three engines ignited, and at T-minus-zero off it went—a beautiful sight. Then, one minute into the flight, *ka-wham*—it looked to Glenn like a nuclear bomb exploding at altitude. Seven astronauts, sobered by the experience of seeing the annihilation of the Atlas, headed off for a few drinks.

The Atlas had been in development since the mid-1950s, but it had still not been tamed and had a failure rate of over 50 percent. This was cause for concern, but with the superpowers duking it out for the first "man in space," the astronauts would fly on it no matter what. It was just a question of how reliable NASA and the air force—which was responsible for the development of the delicate rocket—could make it within a short period of time.

Despite the ongoing drama of the Atlas, things were proceeding at a rapid clip, and it seemed as if the United States was in a good position to send the first man into space . . . until April 12, 1961. On that day, halfway across the world, Yuri Gagarin flew a single orbit in Vostok 1, and any smugness the United States felt about their manned space program went off a cliff. The Western press trumpeted the Soviet achievement in newspapers and on TV and radio to all who could hear: The Russians had sent the first human into space.

To add to the frustration, the Mercury spacecraft had been ready for flight for some time, as was the Redstone. The astronauts were ready too, and Al Shepard had been in training with Glenn as his backup for the first flight. But Wernher von Braun, the man in charge of the US rockets, wanted more tests of the Mercury-Redstone combo, and this had eaten up time. "We could have been first," they thought. Now it was a matter of catching up with the Soviet Union . . . again.

On May 5, just a few weeks later, Shepard rocketed skyward in his Mercury spacecraft. It was a short, suborbital flight—just about fifteen minutes from start to finish compared to the 108 minutes the Soviets had accomplished, but it was enough; America had launched its first man into space, if not into orbit.

Then President Kennedy, with little warning outside the upper levels of NASA and a close circle of administration advisors, dropped a bombshell. Standing in front of Congress on May 25, just a few weeks after Shepard's flight, he put the nation on notice toward the end of an otherwise relatively routine address to the assembled politicos:

I believe that this nation should commit itself to achieving the goal, before this decade is out, of landing a man on the moon and returning him safely to the earth.[29]

**Figure 2.5. Glenn (*left*), with Gus Grissom (*center*) and Al Shepard (*right*). These were the first three astronauts to fly in America's space program. (Courtesy of NASA.)**

He continued speaking for a few more minutes about how this could be achieved, but his intentions were clear: with just fifteen minutes of human spaceflight, America was headed to the moon. It was a galvanizing moment.

This was followed by another suborbital Mercury flight on July 21, 1961, with Gus Grissom at the controls. Glenn was getting impatient; he had assumed he would be first in line for at least one of the early flights but was enduring months as a bridesmaid to the Shepard and Grissom. It was galling—he knew that all seven of them were supremely qualified, but he also felt that somehow his "clean marine" tendencies had worked against him. Many months before, he had chastised the other six astronauts, warning them about the dalliances some were having with single young ladies who were hanging around the bars the astronauts frequented after training, and suggesting that the repercussions of this would be a black eye to the American space program. He was in turn roundly told off by the others, who said that it was none of his business what they did in their off-duty time. Glenn felt that this act may have cost him the opportunity to be first—this was a brotherhood of pilots, and his puritanical tendencies were not popular. And with von Braun and others wanting more suborbital flights to further test the hardware, Glenn saw himself being sent up on just another suborbital hop before the big one—the first orbital flight of the US program.

He could not have been more wrong.

On August 6, the Soviets launched another man into space—Gherman Titov—and this time the Vostok made seventeen orbits. Enough was enough, and even the conservative engineers within NASA agreed that two suborbital flights of the Mercury-Redstone combination were plenty. NASA decided that the next Mercury mission would be into orbit atop the still-worrisome Atlas, and Glenn would be in the cockpit. The launch was set for January 16, 1962.

Glenn started spending entire days in the procedures trainer, making sure that he had every element of the flight down pat. He endured seventy full simulations with over two hundred problems thrown at him, and he was so busy that it barely registered when NASA delayed the launch until January 23. One thing that preoccupied a small part of his mind was photography. He really, really felt that he should take a camera up for this first orbital flight of a NASA astronaut, but the mission planners felt that a camera would be a distraction and might negatively impact the mission. Along with all the other responsibilities the astronaut was assigned, what might happen if a human was tasked with the extra chore of photography? Glenn thought it was absurd and, true to form, took his complaint to upper management.

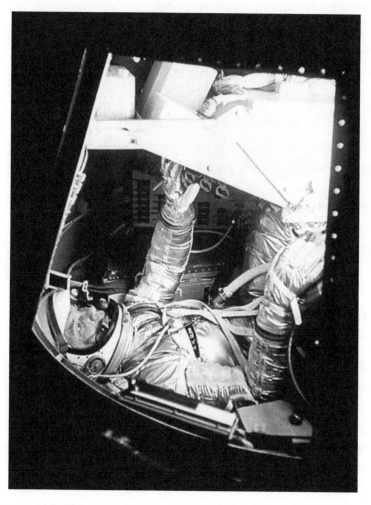

**Figure 2.6. Glenn in the Mercury simulator, 1962. (Courtesy of NASA.)**

Glenn went to Bob Gilruth, the man in charge of manned space-flight at NASA. He got right to the point: "This is ridiculous," he said. "I need to take some pictures, because people are going to want to see what it looks like to be an astronaut. I'm not going to let my own safety and running the spacecraft take second place."[30] Gilruth agreed, and Glenn would get his camera.

**Figure 2.7. Glenn poses with his Mercury spacecraft, Friend-ship 7, in 1962. (Courtesy of NASA.)**

Everything was ready, but bad weather was causing continued delays of the launch. The television networks had set up crews at the Cape to televise the event, and they were as impatient as anyone to see the rocket go. With nothing to show, the news shifted from "Today is the day" to "Will today be the day?"—it was almost like a game of media poker to see who could predict which launch date might be the right one.

On January 27 the weather improved, and Glenn suited up in the predawn hours. Soon after he was riding the elevator up the launch tower and was sealed inside the tiny Mercury capsule. The hatch was bolted closed, leaving him with just enough room to move his arms inside, the control panel just a couple of feet ahead of him. And there he sat for the next six hours; by midday it was clear they would not be launching that day, and the crew unbolted the hatch and sent Glenn packing.

**Figure 2.8. Glenn mugs with the supervisor of launchpad operations, Gunther Wendt, after a scrubbed launch in 1962. (Courtesy of NASA.)**

Glenn waited out the weeks before another launch would be scheduled, and then finally, on February 20, it was time to go. It was the eleventh scheduled launch date—maybe this would be the one? Glenn arose at 1:30 a.m. and went to the dining room for breakfast. It was just the third manned flight for NASA, but the launch day breakfast was already a tradition—a selection of food that would be filling, provide energy for the demanding mission ahead, and was "low residue"—

that is, would be less likely than other foods to require a bowel movement in the next few hours. There was no facility for pooping inside the Mercury capsule—in fact, there was not even enough room inside to accommodate such needs. Glenn ate his steak and eggs with some NASA officials and Deke Slayton. The weather was iffy, maybe fifty-fifty for a launch, but they would try once again.

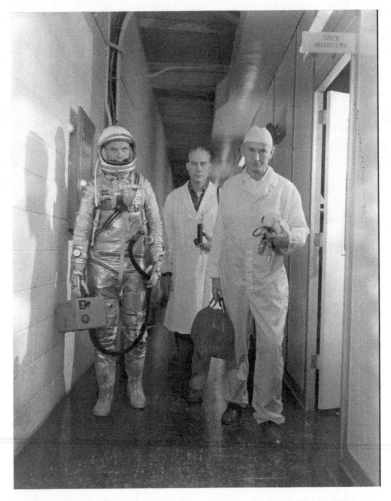

**Figure 2.9. John Glenn, suited up and carrying his portable air conditioning unit, walks to the launchpad with technicians in 1962. (Courtesy of NASA.)**

Soon he was suited up and had ridden the transfer van to Launch Complex 34. It was still dark, and the Atlas stood there looking like a silver arrow aimed at the black sky. By the time the sun had risen, Glenn had taken the elevator up to the capsule, some eighty-five feet above the pavement. The technicians soon had him ensconced inside the spacecraft and the hatch bolted on once again.

As Glenn waited, he went through the checklist. Then he waited some more. He chatted with Scott Carpenter over the radio, then with his family for a bit, while he waited out the delays. He finally signed off with his wife by saying, "Hey, honey, don't be scared. . . . I'm just going down to the corner store to get a pack of gum."[31] It was his stock line that he'd used with Annie since the war years. *No need to fret, it's no big deal.* She was used to it by now. Each of them professed their love, and then it was back to business.

The thin-skinned Atlas creaked and groaned beneath him—the ultra-cold cryogenic liquid oxygen was being topped off, and the metal in the pipes and fittings was voluble in its protest. Soon the engines would ignite, digesting a ton of fuel per second, and 341,000 pounds of thrust would be pushing against all that thin stainless steel; Glenn hoped the booster didn't crumple like a soda can. The rocket had been modified and strengthened since those early failures (had it really only been a couple of years?), and Glenn was confident in the machine, though he'd had to force himself to push the memories of the Atlas' launch failures aside when he first saw the rocket standing in the predawn light.

The countdown continued without additional holds now. With eighteen seconds left before ignition, control switched to the rocket's onboard sequencers—that's when it all became sank in: he was finally going to fly, and this one would be into orbit.

At 9:47 Eastern Time, the engines ignited, seconds later the hold-down clamps released, and he was off. The rocket started slowly but picked up speed as the fuel tanks emptied—the g-forces maxed out at about eight times normal but lessened as the rocket neared orbit.

At just over five minutes into the mission, the last of the Atlas' three engines shut down and the Mercury capsule popped free of the booster. John Glenn was in orbit, and with him rode the hopes of his country. He had a moment of great pride when he radioed down, "Zero-g and I feel fine . . ."[32] Then, a moment later and with elation, "Oh! That view is tremendous!"

Within moments the capsule had rotated to fly heat shield forward, and Glenn could see the Atlas slowly tumbling behind him, engines spent and its work done. CAPCOM (capsule communicator) Al Shepard told Glenn that he was go for at least seven orbits, though the planned mission was for just three. That assumed that everything continued to work properly, of course.

As he flew over the Atlantic, nearing Spain, Glenn tested the maneuvering system. While the Mercury capsule was designed to hold its orientation automatically, the astronaut had the ability to directly control the small thrusters that could change the capsule's orientation if needed. The system checked out.

Glenn then unpacked the camera he had worked so hard to get included on the mission—it was a small Minolta 35mm compact. NASA had looked at expensive Leicas and other cameras, but none could be properly modified for use during the flight. But Glenn had recently wandered into a drug store near the Cape, and there it was: a Minolta 35mm camera with automatic aperture control. He paid $45 on the spot, for which he was never reimbursed. It must have been the cheapest component of the space program.

Grabbing the lightly modified camera, Glenn snapped some shots as his hundred-mile-high orbit took him toward Africa.

As he crossed the eastern coast of Africa, Glenn did some light exercise (as much as could be done in the tiny spacecraft) and recorded medical readings for the doctors in NASA's tracking station in Zanzibar, which would be relayed to the Cape. "We wanted to try out some exercise equipment," Glenn recalls, "So we had bungee cord there. And I was to use that prior to taking my blood pressure a couple of times, so we could see what the effect was from exercise up there."[33] He would repeat the medical tests—basically blood pressure readings, eye tests, and other things that were not being tracked by the biomedical sensors stuck on—and in—his body. NASA wanted to know as much as possible about the physiological condition of its astronauts in flight. Everything checked out fine.

As he flew on, now about forty-five minutes into his mission, he entered the Earth's shadow—the orbital night. The quick sunset was spectacular, and once the sky was black he spent some time identifying constellations. This was important research for later missions—Apollo specifically—that would be making navigational fixes by using celestial navigation.

As he approached the coast of Australia, Glenn could see the glow of lights below him. Fellow astronaut Gordon Cooper was CAPCOM at the tracking station there, and chatted amicably with Glenn as he passed over. "That was about the shortest day I've ever run into," Glenn replied. "Kinda passes rapidly huh?" drawled Cooper.

Glenn spotted the Pleiades star cluster, also known as the Seven Sisters, and marveled at its natural beauty. Then his eyes moved back to the lights below. Cooper had mentioned that Glenn might notice some lights below him, and did he ever. "I can see the outline of a town, and a very bright light just to the south of it," Glenn said.

Cooper said, "[That's] Perth and Rockingham you're seeing there." The residents had made a point to leave their lights on late, and many went outside to wave flashlights as Glenn passed over. He was deeply touched by their efforts. "Roger, the lights show up very well," he said enthusiastically, "and thank everyone for turning them on, will you?" Cooper replied, "We sure will, John."

Glenn drifted on toward Hawaii, and had his first meal—a tube of applesauce. It was nothing special—just plain applesauce from a toothpaste tube—but the doctors wanted to make sure he could swallow properly in zero-g, even though there was plenty of evidence that the Soviet cosmonauts had done just fine on their increasingly longer flights.

Glenn was just over an hour into the mission, and everything was going well as he drifted into his first orbital sunrise. He was a huge fan of sunsets and sunrises and often said that he "collected them" in his mind—to him, they were some of God's greatest signs to humanity. Then something else grabbed his attention.

He radioed down to the nearest tracking station on Canton Island, a flyspeck atoll in the vast mid-Pacific. "I'm in a big mass of thousands of very small particles that are brilliantly lit up like they're luminescent," he said. "They are bright yellowish-green. About the size and intensity of a firefly on a real dark night. I've never seen anything like it."[34]

He went on to describe the apparition in great detail when he came into range of the next tracking station at Guaymas, Mexico:

> They swirl around the capsule and go in front of the window and they're all brilliantly lighted. They probably average maybe 7 or 8 feet apart, but I can see them all down below me, also. ... They're very slow; they're not going away from me more than maybe 3 or 4 miles per hour. They're going at the same speed I am approximately. They're only very slightly under my speed. Over. They do, they do have a different motion, though, from me because they swirl around the capsule and then depart back the way I am looking.[35]

Glenn was surprised at the cool reaction on the ground—the CAPCOM simply continued to read off retrofire settings for his upcoming reentry. What he did not know was that when his reports reached Mercury Control people there started to worry. Was there something wrong with the spacecraft? Was something falling apart and leaving luminescent fragments floating around the capsule? Some posited that the heat shield might be disintegrating in some unknown fashion—even though the shield had been thoroughly tested, they thought they might have missed something.

But Glenn did not have long to dwell on the "fireflies." A new wrinkle grabbed his attention. A thruster started firing—odd, since the spacecraft seemed to be in the proper orientation, and there was no apparent reason that the automatic guidance system should be causing the thruster to fire. Then another thruster fired to correct the motion. Maybe it was a one-time thing, he thought. But when it happened again, Glenn took over with manual fly-by-wire, commanding the spacecraft with a hand controller. Then he switched back to automatic, and the thrusters started popping again. Glenn radioed Mercury Control and they agreed that he should keep the system on manual control for now, lest he run down his maneuvering fuel reserves.

As he went into his second orbit, he noticed that one of the thrusters had stopped working altogether. It was not urgent—there were two on each axis, one large and one small, but this was definitely something that he would note for his post-flight briefings.

Then, as he passed over the Indian Ocean, a puzzling call was relayed from Mercury Control. "Keep your landing bag switch in off position," they said. He noted that it already was, and he knew there was no reason to change it at this point in the flight. The landing bag was emplaced between the heat shield and the spacecraft, and was intended to be deployed just prior to splashdown, to act as a cushion when the capsule impacted the ocean.

As he was puzzling over the message, he noticed that the capsule had drifted off axis again. Testing the maneuvering system in the automatic setting caused the spacecraft to pitch and yaw in multiple axes. A visual check of his gyro system—the spinning devices that told the spacecraft what direction it was pointing—indicated that *it* thought he was in the proper orientation. But his eyes told him a different story.

"I have some problems here with ASCS [his attitude control system]. My attitudes are not matching what I see out the window." His fuel was down to about 60 percent, and with the flight just over the halfway point of the minimum three orbits planned, he switched once again to manual control to conserve fuel.

Then, as he flew over Australia again, Cooper came back on the radio, asking to confirm that the landing bag switch was in the off position. This was getting irritating.

"That is affirmative. Landing bag switch is in the center off position."

Cooper then asked if Glenn had heard any banging noises when maneuvering the spacecraft, and Glenn said no. "They wanted this answer," Cooper said cryptically.

Glenn was a pilot, and a marine, and now an astronaut. You followed orders without question in the first two of those disciplines, but as an astronaut he felt that he deserved some indication of what might be worrying them—but Cooper said nothing more about it.

Then over the next tracking station he got a bit more of an indication of what had the ground controllers concerned. "We also have no indication that your landing bag might be deployed," they said. Aha. Maybe they were worried that the "fireflies" he had reported before, and again a few minutes prior, had been a result of a premature deployment of the landing bag? But no more information was forthcoming, and he waited them out.

Glenn flew into daylight again, and into his third orbit. He went back to automatic attitude control again to test the system, with the same results—the attitude indicator thought it was right on the money, but he could see that the spacecraft was off-course. Back to manual control.

As he glided over Hawaii, Glenn finally got an answer to the riddle of the landing bag. The CAPCOM there said, "We have been reading an indication on the ground of segment fifty-one, which is the landing bag deploy. We suggest this is an erroneous signal. However, Cape would like you to check this by putting the landing bag switch in 'auto' position and see if you get a light."

Now Glenn understood the gravity of the situation—his worst fears confirmed. If the landing bag had deployed, that meant that the heat shield was loose, flopping around on a few feet of inflatable cushion. If the shield slipped to one side during reentry, this would allow the searing heat from friction with the atmosphere enter the bottom of the capsule, and this would incinerate him. Likewise, even if the shield was just slightly off axis, it could cause the capsule to rotate and, again, there would be no protection from the heat shield. Either condition would deliver fatal results.

"I remember thinking then that if the heat shield was detached some way and was not working right, [that] the first place you're going to feel heat is on your back, and if you did . . . it wasn't going to be very long," he later said.[36]

But on the Mercury capsule, there was another line of defense: the retrorocket pack. Inside this unit, which was strapped to the heat shield, were three small rockets, designed to slow the spacecraft enough to reenter. As designed, the retrorockets would fire, and then the package containing them, called a retropack, would be discarded by firing small explosive charges that cut loose three metal retaining straps, and the retropack would drift away before the capsule plunged into the atmosphere.

Glenn moved the landing bag switch into "auto" as requested, and the light did not come on—everything appeared to be fine. But there was still an indicator light down at the control center telling them that the bag had deployed. With the troubling questions about the landing bag in mind, the flight director decided to end the mission at three orbits, the minimum they had planned for.

With just minutes to go before the retrorockets fired, the voice of Wally Schirra, the CAPCOM on the US west coast, came through the earphones: "John, leave your retrorocket pack on through your pass over Texas," he said. Glenn concurred, and the first of the three retro-rockets fired right on time. Then the second, and the third. John Glenn was coming home, as either a triumphant first-American-in-orbit, or as a glowing cloud of incandescent cinders.

As he flew over Texas with his altitude starting to drop, Glenn got another message: "We are recommending that you leave the retro package on through the entire reentry." Glenn acknowledged that he understood. He again asked why—he wanted to hear them say it—but got no answer.

Then, finally, as he was passing over Florida for the last time, Al Shepard, who was manning the console at the control center, said, "We are not sure whether or not your landing bag has deployed. We feel it is possible for you to reenter with the retro package on. We see no dif-ficulty at this time for that kind of reentry."

Finally. It had taken one of his own—another astronaut—to relay the news. While Glenn did not know it at the time, there had been a furor in Mission Control over how much to tell him. The other astronauts felt strongly that as a pilot (and astronaut), Glenn should know *exactly* what they knew on the ground about the condition of his spacecraft. Others, the doctors among them, were concerned that he might panic. Shepard and the others knew that this was absurd—they had all faced worse in combat. The argument raged on for much of Glenn's flight and would not be resolved until his return and debriefing, during which he made his feelings clear: from then on, NASA would tell their astronauts *everything* they knew, when they knew it. End of discussion.

As Glenn deorbited, one more partial recommendation started to come in from Shepard, then ... static. Glenn had entered the Earth's atmosphere, and the ionizing effects of the air, now heating up his heat shield, blocked all radio communication. He would be alone, plunging through a blazing envelope of fire, until shortly before splashdown.

The capsule shuddered and shook as he followed a curving line to the Atlantic below. To Glenn's relief, the heat shield seemed to be doing its job as planned. As he later recalled, "Your heat during re-entry on the heat shield was around 3,000 degrees. . . . Out about two and a half or three feet in front where the plasma layer is, it gets up around 9,000 degrees, which gets close to the surface heat of the sun. Thus it is very important that your heat shield be in place."[37] That's an understatement worthy of a test pilot.

The heat shield had been tested, both on the ground and in flight, but Glenn was the first American to trust his life to the designers of the Mercury capsule in reentry. Schirra had earlier made a now-common joke while talking to Glenn a short while earlier about the components being built "by the lowest bidder," and while it was worth a chuckle at the time, it was not quite as funny now.

Down, down, down he went. A bright yellow-orange cloud enveloped the capsule and was all he could see outside his window. Then there was a thudding noise, and a few bits of white-hot metal flew past the window—the retropack straps had melted and let go. Now it was up to fate, and in Glenn's mind he was in the hands of God.

As he passed 45,000 feet, he tried the hand controller again but then realized he was out of fuel. No matter, the Atlantic Ocean, vast and wide, was waiting for him no matter what he did now. At 25,000 feet a small parachute, called a drogue chute, opened to stabilize the falling spacecraft, then, 15,000 feet later, the larger main chute was released. Within moments, he felt a hard thump, and Glenn was floating in the ocean awaiting the navy recovery forces.

The flight had been a tiring experience, and one with its share of problems, but he felt like a million bucks.

The navy delivered him to Grand Turk Island in the Caribbean, where some of the other astronauts were waiting … along with a gaggle of NASA doctors, engineers, and technicians. The debriefing was arduous but necessary—they wanted to get his impressions of the flight recorded as quickly as possible. This process continued when he got back to the Cape. Other than his stern words about letting an astronaut know what was going on with his spacecraft during flight, Glenn accommodated them with patience.

Then came chaos. NASA had succeeded in putting their first man into orbit, and if the hoopla surrounding Alan Shepard's flight had been intense, the reception Glenn received was exponentially more so. As soon as he was back in Florida, Glenn was reunited with his family, and then they were driven to an impromptu parade in Coco Beach. For twenty miles, people lined the sides of the road, applauding, waving, and throwing kisses. Glenn took it in good natured stride, but it was just a foretaste of what was to come.

Soon Glenn was in Washington, briefing the president on his adventure, and was then loaded along with his wife into another convertible for a parade down Pennsylvania Avenue. It dwarfed the Coco Beach event, the streets clogged with citizens of all stripes—cheering, waving flags, crying. Glenn and Annie were overwhelmed, as was NASA's public relations team. They had expected public enthusiasm—that had been made clear by Shepard's flight—but this was simply off the scale.

The procession ended at the Capitol Building, where Congress had been assembled to hear from America's newest space sensation. Glenn did what Glenn did best—he came off as the true, humble, American hero. True to form, he closed with, "As our knowledge of the universe in which we live increases, may God grant us the wisdom and guidance to use it wisely."[38] He brought them to their feet with a standing ovation.

COCOA BEACH PARADE

**Figure 2.10. Glenn rides with President Kennedy for the impromptu Cocoa Beach parade, 1962. (Courtesy of NASA.)**

Next came the obligatory open-car parade through the streets of New York. But Glenn made one thing clear to NASA: the astronauts had done this as a *team*. This was not a parade for John Glenn, but for all the guys who had—and would—put their lives on the line. NASA acquiesced, and Glenn's comrades joined him for the event. The entire city turned out, it seemed, and all previous records were broken in the one metric that was unassailable: the sheer tonnage of paper showered on the astronaut procession—ticker tape, torn up letters and newspapers, and bits of telephone books—outweighed that from any previous event. It had been launched into the frigid March air over New York by over four million people.

Glenn then returned to his home town, New Concord, Ohio, to be feted there. Another city, another parade. But this one was even more surreal. "New Concord normally had a population of about 1,100 and about another 1,000 students in college. So we had about 50,000 people into a 1,500 people town. That was quite an experience," Glenn recalled.[39]

**Figure 2.11. John Glenn with Annie after his Mercury flight. (Courtesy of NASA.)**

And then, somewhat predictably, some powerful people began to speculate on Glenn's possible future in politics. James Reston, a *New York Times* columnist, wrote that Glenn embodied "the noblest of human qualities."[40] The Republican Party suggested that he run for president. Others suggested Congress—among them Glenn's newfound friend Bobby Kennedy. For the time being, Glenn was determined to remain in the astronaut corps—but this idea soon turned stale.

Glenn continued to work for NASA, but he alternated between being anchored to a desk and trotted out for public relations duties, increasingly being used to justify the cost of the program to politicians and the public. All the while, the rest of the Mercury 7 made their record-breaking flights. Glenn was happy to support the program, but he increasingly felt the itch for another flight assignment. He waited, and waited. The Mercury program ended in 1963, and in 1965, the Gemini program was scheduling astronauts to fly in a new, improved two-seat spacecraft with greatly enhanced capabilities—but Glenn's sense was increasingly that there would not be a seat for him. It was time to move on.

In January 1964, he announced his candidacy for a seat in the US Senate as the representative from Ohio. But this did not last long—a month into the campaign, Glenn slipped in his bathroom and hit his head—hard. He experienced severe vertigo for months and was barely able to stand, much less press the flesh on the campaign trail. Within two months he announced his withdrawal from the campaign—"I do not want to run as just a well-known name," he said.[41]

By 1965, Glenn had retired from both NASA and the Marine Corps—what was next for the first American into orbit? After considering a number of offers, Glenn worked as an executive with Royal Crown Cola, a soft drink company. He enjoyed the challenges of business, continued to stump for NASA in public relations appearances, and expanded his influence into promoting the Boy Scouts of America and other nonprofits.

After five years in corporate life, Glenn took another run at the Senate but lost in a close primary—he was grossly underfunded and was outspent by four-to-one in the campaign. Then, in 1974, he gave it another shot and won his Senate seat. He was reelected in 1980 with the largest margin in Ohio's political history.[42] Glenn aimed for the presidency in 1984, but it was a short-lived campaign and he pulled out early. He returned to the Senate in 1986 and again in 1992, for his fourth term. Among other important issues, Glenn aggressively tackled nuclear nonproliferation.

During his tenure in the Senate, Glenn had been a staunch defender of the NASA budget—it was a way to stay engaged with the space agency, but in his heart he still yearned to venture into space again. The space shuttle had been flying since 1981, and other than the shutdown experienced after the loss of the *Challenger* in 1986, was heading into space frequently, for increasing durations. By this time NASA was looking at the effects of prolonged weightlessness on humans, and Glenn paid close attention to the studies. It appeared to him that these effects—loss of bone density, changes in eyesight, and other maladies—were not dissimilar to what occurred as normal people aged. He lobbied NASA to do some experiments in space along these lines with their frequent space shuttle flights. Then he tried to sharpen the deal—he volunteered to go along as the subject of the study.

**Figure 2.12. John Glenn with his wife, Annie, in 1965. (Courtesy of NASA.)**

His entreaties were greeted with polite smiles and agreeable words fitting for a seventy-year-old former astronaut. He was not taken seriously, even at the top—he had met with then NASA administrator Daniel Goldin, and made his pitch, but he got the requisite polite grin and thanks for his interest.

Glenn was reaching the end of his career in the Senate, having decided in 1997 that he would step down at the end of that term. For the past number of years he had still harbored his dream of flying on the shuttle, and had been scheduling regular medical tests to assure his fitness and overall health—just in case. Then, in January 1998, he got a call from NASA—it was Goldin. He said, "You're the most persistent man I've ever met."[43] Goldin said that Glenn's physicals had checked out, and that he had decided to send him back into space after all.

Glenn was assigned to STS-95, a 1998 flight of the shuttle *Discovery*. He would not be piloting this time—that was reserved for younger, newer astronauts, who had flown nothing but the shuttle. He would be a payload specialist, a general category for the non-piloting shuttle astronauts, and also the subject of intensive medical testing. He was now seventy-seven and would be the oldest person to fly into space—a distinction that remains today—and was an average of twice as old as the other astronauts. It did not bother him one bit.

His schedule was now reminiscent of the early 1960s—back and forth between Houston and Florida for training, with occasional trips home. On October 29, 1998, preparations and training complete, *Discovery* hurtled into orbit. For the next nine days, Glenn participated in medical experiments as both the researcher and the test subject. He was also assigned to perform duties related to photography and videography during the flight. He suffered no ill effects, and the research yielded what he felt were important results, though he later expressed disappointment that these studies were not followed up adequately.

Upon *Discovery*'s return, in another bit of déjà vu, there was a short ticker tape parade down the main street leading out of the Johnson Space Center. It was not quite the same as New York in 1962—people were tossing streamers and confetti from the sidewalks as he was slowly driven down the thoroughfare, not from skyscrapers, since the area surrounding the space center is comprised of the same flatlands that make up the once swampy region in which the NASA facility sits. But nobody minded the more modest scale of the proceedings—John Glenn, the seminal American hero, had returned safely home from space once again.

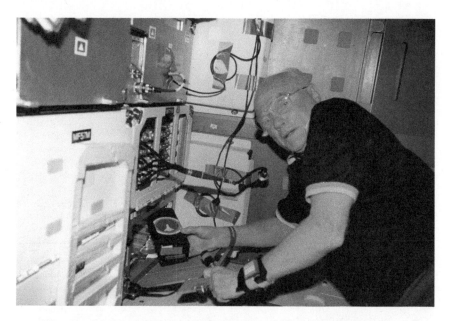

**Figure 2.13. Glenn during the shuttle mission of STS-95, performing an experiment in osteoporosis (bone density). (Courtesy of NASA.)**

Shortly after landing, Glenn summarized his feelings about his second experience in space by saying, "To look out at this kind of creation and not believe in God to me is impossible."[44]

Glenn's final term in the Senate ended in January 1999, and he was now finally retired from the many aspects of public life that he had undertaken in his long and impressive career.

John Glenn died on December 8, 2016, at age ninety-five. Annie, his lifelong sweetheart and wife of seventy-three years, was at his side. He had taken his final stroll to "pick up a pack of gum."

Figure 2.14. Glenn walking with NASA administrator Dan Goldin (*left*), after his flight aboard the space shuttle. (Courtesy of NASA.)

# CHAPTER 3

# VALENTINA TERESHKOVA: FLIGHT OF THE SEAGULL

**Figure 3.1. Valentina Tereshkova during training in 1962.**

Yuri Gagarin called her *Chaika*, Russian for seagull. It seemed a fitting appellation for a young woman in her mid-twenties, a gymnast and daredevil who had grown up during some of the Soviet Union's most challenging times after World War Two. Before she was selected for spaceflight, Valentina Vladimirovna Tereshkova had been a simple girl, a young textile worker in a clothing factory, and had little formal education. Other than her passion for skydiving, she was just like millions of other young Soviet women. She could never have dreamed that she would have a nickname bestowed upon her by the most famous man in the Soviet Union while she prepared to fly in outer space.

When Gagarin made the first flight of a human into space in 1961, it was a triumph. The space race had been aflame since 1957 and the flight of Sputnik, and besides launching ever more capable satellites, the next Soviet goal was putting a human into space. Gagarin's single orbit had clinched that "win" for the Russians, once again serving to secure their propaganda efforts at being the more technologically advanced superpower. But what spaceflight first should come next to upstage the West?

Flying multiple cosmonauts in a single spacecraft would be a record-breaker, as would a rendezvous of two spacecraft in orbit. Longer duration flights were also a worthy goal, and were in the works. A spacewalk—a cosmonaut leaving the safety of their spacecraft and floating free in space—would also impress the world. But these were all in the future—the technology was not quite there yet, though it would be in short order. What might the Soviets do in space that the West would not be attempting with their own Mercury program, now running full bore? Whatever it was, it would have to be done using the existing Vostok single-seat spacecraft.

Nikita Khrushchev, the leader of the Soviet Union, had an idea. Why not put a woman into space? It was a no-brainer to him—this would definitely be a first, since as far as anyone knew the Americans were not seriously considering a woman for spaceflight (in fact, they were—thirteen women were quietly undergoing a form of training for a possible Mercury flight, but that program did not reach fruition). Putting a woman into space would also serve another purpose, a political one: in Soviet ideology, men and women were supposed to be true equals (though there was still a firm foundation of chauvinistic thinking in the Soviet Union just as there was in the West), and what better demonstration of this than a female cosmonaut? A woman cosmonaut would be a propaganda coup, and would provide another set of data points for human physiology in spaceflight to boot.

Sergei Korolev, the chief designer of the Soviet space program, was already planning a long duration flight. Khrushchev agreed to this on one condition: the first flight of a woman would have to coincide with a spaceflight duration record and a two-spacecraft rendezvous. This would net three firsts in one—a true space spectacular. Korolev agreed, and the die was cast.

There were no women in the cosmonaut corps—it had never before been a serious consideration. They would have to work fast to select someone and begin training. There were few female jet pilots, and since the Soviet cosmonaut requirements were not dissimilar from their American counterparts—they wanted test pilots—that meant no ready pool of candidates. And there was an additional caveat: Khrushchev, true to his Communist roots, wanted a "regular girl," a member of the proletariat working class; someone that would represent the flower of Soviet womanhood.

As with many historical records of the Soviet Union in the space race era, there are various versions of this story that do not always agree. One suggests that the germ of the idea to fly women did not start with Khrushchev, but occurred when Soviet cosmonaut Gherman Titov and the chief of their cosmonaut corps, Nikolai Kamanin, were visiting John Glenn for a barbecue at his home in 1962. As the tale goes, Glenn was talking proudly about the "Mercury 13," the women being trained for a possible flight in the Mercury program, and that one might fly in Mercury before then end of the year. The Soviet guests returned home and told their superiors that the Americans might beat the USSR by sending a woman to space, which may have led to an acceleration of Soviet plans. Other sources indicate that the Soviets had recruited their female cosmonaut candidates months before this barbecue occurred. Sadly, as noted, none of the women trained for Mercury ever flew—in fact, no American woman flew into space until the shuttle era, an oversight that has thankfully been corrected since then by many, many flights of highly competent and successful female astronauts.

Once Khrushchev and Korolev had made their deal, Yuri Gagarin was tapped to lead the selection process, which began in mid-1961. There was little time to waste—the fight was to occur in less than two years, and although the Vostok was essentially automated, whoever sat in the pilot's seat would have to be as calm and unflappable as the other cosmonauts—human spaceflight was still a new and very dicey business.

Tereshkova was born in 1937, in a village called Maslennikovo in central Russia, about 160 miles northeast of Moscow. The village was part of a farm collective, and the forty structures had neither running water nor electricity—the conditions were spare. Her father had been a tractor driver before serving in World War Two and her mother a textile worker; her mother's vocation had guided Tereshkova into textile factory work after a stint working at a tire factory. On the political side, her family had no blemishes—her father had died early in the war, and young Valentina had not only done well in her labors but was also the secretary of the local Young Communist League. She was also an apparently fearless girl, what we might have called a "tomboy" in the West, riding horses bareback, swimming across the wide, chilly Volga River, and later pursuing skydiving, jumping dozens of times with war-surplus parachutes. Notably, the other three candidates were also skydivers; a fairly elite set of Soviet women in 1961.

The politicians might have had a different opinion had they known Tereshkova's somewhat more complex history with daredevil sports. While she was very athletic, and had swum the Volga multiple times, skydiving terrified her at first. She was compelled to attend a meeting of the local parachute club by a friend but was put off by what she saw. This was a sport for men, for adventurous souls. She was just a girl, and not suited to such things, she thought. Watching the tiny figures tumble out of an airplane far overhead finalized her decision—this was not for her.

Later, at Weaving Factory Number Two where she worked, activity clubs were being organized and workers were encouraged to join them. Activities included boating, amateur radio, target shooting ...

and skydiving. Upon this second exposure, and perhaps with a slight numbing over time, she decided to give it a try. She was just twenty-two years old.

After a period of instruction on the ground, she went up for her first jump. The old Yak-12 airplane was noisy, with the roar of the motor and the wind ripping past the open door as it flew. Tereshkova was surrounded by people who she thought all looked calmer and more prepared than she felt, and a stoic instructor was seated across from her. She was fearful of the jump but more afraid of screwing up—she thought she might freeze when the time came.

After some rough air, the plane settled into smooth, level flight. Tereshkova was keyed up—this must be it! But the air roaring past the open door was so loud she was afraid she might not hear the jump master's instructions to leave the plane. She waited, and there it was! Time to go! She stood up, bolted for the open door, and out she went.

Unfortunately, the slack-jawed instructor had said nothing, and could only watch as his first-time trainee abandoned the plane.

She reached the ground, rolled as she had been taught to do, and stood up—that had been glorious! It reminded her of the high jumps she had made into a nearby river when she was younger, except this had been much more thrilling—she wanted to skydive every day. As she was gathering her parachute, another instructor strode up to her and asked her why the hell she had jumped without being told to? She was flabbergasted and chagrined, but the other students just laughed it off, and the instructor broke a smile and told her that she'd done alright for a beginner, that she would learn. Next time, just follow instructions.

After this premature beginning, she fell in love with the sport and skydiving anytime she could. She improved quickly and began to enter competitions for precision parachuting. The daredevil in Valentina had been awakened to a new form of adventure.

Unfortunately, her mother was not so enthused. She hated the idea of her little Russian doll jumping out of airplanes—not surprising for a woman who, like so many, had lost her husband to the violence of war.

Losing a daughter would be unthinkable. But Tereshkova persevered, and ultimately made hundreds of jumps.

**Figure 3.2. Tereshkova was an avid amateur parachutist in her early twenties.**

On April 12, 1962, Tereshkova had just started a political meeting at the textile factory when she heard a ruckus outside the room—cries of triumph echoed down the halls. A Russian had orbited the Earth! She and the rest of the employees were stunned and thrilled. With no pre-launch fanfare or prior announcement, Yuri Gagarin had beaten the Americans into space. What an amazing thing, she thought.

"When we learned that Yuri Gagarin had landed safely our joy knew no bounds," she later said.[1] "All of the town's inhabitants were out in the streets." She thought to herself, "First the men will fly, then we women," but she did not really believe it would happen anytime soon. She certainly had no idea that she would be the one at the controls. Flying into space was something she would love to do, but who would put a factory laborer of common roots into a spacecraft? Cosmonauts were elite military men, usually well-connected politically, and certainly smarter and stronger than her. But then she read that Gagarin had started out not so differently from herself.

"I read in a newspaper that Gagarin was a student of an aerospace club, just like I was, and then I decided. I'll be [a] cosmonaut," she said.[2]

She wrote to the Soviet space agency, and her letter went into storage with thousands of others. It was supposedly unearthed when the search began for suitable candidates, although it's hard to know if this really occurred, or if it was just good propaganda to publish after her flight. In any event, she was soon quietly selected to join the first and only group of women to compete for the coveted single women's flight slot in the Vostok program. Upon further evaluation, the four hundred or so candidates were soon narrowed down to a group of fifty-eight, and then to just four. They all had backgrounds in aviation or skydiving clubs.

Under a shroud of secrecy, training for the four female cosmonaut candidates began immediately—even Tereshkova's mother did not know what her daughter was doing. Tereshkova told her mother—under orders from the Soviet space authorities—that she was going to Moscow to train with an elite women's skydiving team. The elder Tereshkova would not learn of her daughter's history-making flight until seeing it on the news after the fact, as was the Soviet way. Soviet space accomplishments were only *news* after they had succeeded; otherwise they was given a generic flight designation and put quietly aside.

**Figure 3.3. Tereshkova revered Yuri Gagarin after his flight, but had no idea she would soon be working at his side as a cosmonaut.**

"It was top secret," Tereshkova would later say. "My mother, [like] the mother of Yuri Gagarin, first knew about it with the rest of the country. It was a very big surprise."[3] That's putting it mildly, but her mother, while frightened for her daughter and indignant at first, was of course extremely proud. She heard of the news from friends who saw images of what looked like her daughter in space, and she had to convince herself that this was indeed her daughter Valentina.

Training for the upcoming Vostok flight was rushed and not as in-depth as that of their male counterparts, but was still arduous. More skydiving training was undertaken, interspersed with dozens of hours of classroom work to provide the basics of engineering, science, and navigation, capped by hours of jet flight with an instructor. While these were all essential skills to be a cosmonaut, they weren't as applicable in the automated Vostok, which lessened the need for thorough training in all these areas. Nonetheless, the women persevered, each hoping to be the one chosen for what might be (and was) just a single spaceflight for a woman.

There they were, all together in the same rooms at Star City, performing the same tasks with their heroes. "During classes I often sat near Gagarin and Titov," Tereshkova would later say—she was thrilled yet humbled by being in the presence of the two men. "We were right next to each other, like school kids, at the same desk. I had to do everything they did. . . . [I]t was hard to compete with them. . . . [M]y heart of a common factory girl was beating with happiness!"[4]

The Soviet press had already trumpeted the nation's intention to fly women into space. There might even be two missions flown simultaneously, both with women at the controls. And why not? The government had multiple Vostok spacecraft ready to go, so it was more of a question of who would be at the controls. And two women in space at the same time ... flying in close proximity ... what a public relations coup that would be (for a country that claimed to decry public relations).

Soviet officials were down to the final selection—there were two candidates, both fully qualified. According to records released after the fall of the Soviet Union, the final choice came down to politics above all. When the two finalists were interviewed for final evaluations, Tereshkova cleared the hurdle. Though possibly not the front-runner going in, she left the interviews as the clear winner. When her rival, Valentina Ponomaryova, was asked what she wanted from life, she said, "I want to take everything it can offer"—not an unreasonable response from a young woman sufficiently aggressive to be interested in skydiving in the USSR of the 1960s. But when Tereshkova answered the same question, she was far more canny. Having been involved with the Communist Party for a few years at this point back at the textile mill, she knew exactly what to say: "I want to support irrevocably the Komsomol [a communist-led youth organization] and Communist Party."[5] That, apparently, did the trick.

On a side note, lest you think Tereshkova to have been insincere in order to score points with her examiners, in a book she wrote in the 2000s her tone is irretrievably patriotic—it was her duty to the motherland, all the cosmonauts always treated her fairly and with nothing but respect, and life as a cosmonaut was wonderful. She may believe it, but all historical indications lead to the conclusion that the experience was far more challenging than that for a twenty-six-year-old woman.

Khrushchev himself was the tiebreaker in the decision. Tereshkova was his tough-yet-buttoned-down pure Russian peasant girl. She was the salt of the earth, humble yet determined, a good Communist, and, apparently, virtuous. Ponomaryova had at one point apparently made statements indicating that she felt women should be able to smoke and still retain their virtue (not shocking from today's perspective, given that in the Soviet Union at that time, smoking and hard drinking were both endemic). It was the wrong thing to say for an aspiring female cosmonaut. Finally, there was Kamanin's assessment: the chief of the cosmonauts called Tereshkova "Gagarin in a skirt." Given Gagarin's status as a Soviet (and world) hero, that was high praise indeed.

Tereshkova and her backup, Ponomaryova, continued to train and prepare until shortly before the mission. With the classroom work behind them, it was now about physical toughening and medical tests. While spaceflight appeared to be safe for men, the Soviet scientists and doctors wanted to make sure the first (and, for nearly twenty years, only) woman in space would survive and perform well. Already at her prime at twenty-six, and an athlete before she started in the program, she may have been one of the strongest women in the USSR by the time she flew. She was certainly the best prepared, and had by then been commissioned as a junior lieutenant in the Soviet Air Force.

And she certainly played the part of the proud Communist well. At one point close to the flight date, she was quoted as saying, "Since 1917 Soviet women have had the same prerogatives and rights as men. They share the same tasks. They are workers, navigators, chemists, aviators, engineers. And now the nation has selected me for the honor of being a cosmonaut. As you can see, on earth, at sea and in the sky, Soviet women are the equal of men."[6] These were certainly good ideas, but were before their time. While the Communist Party espoused these ideals, the reality was somewhat different, even if she did not know it at the time (though she probably did).

**Figure 3.4. Soviet leader Nikita Khrushchev chats with Tereshkova during a public ceremony after her historic flight. (Courtesy of Wikimedia Creative Commons, RIA Novosti archive, image #159271 / V. Malyshev / CC BY-SA 3.0.)**

At one point in the mission planning, Tereshkova was scheduled to launch in Vostok 5, and after she reached orbit, Ponomaryova would follow in Vostok 6. The two women would be in orbit simultaneously—beat that, America! But this plan was changed just months before the scheduled launch date. The flight assignments were reversed, with Tereshkova slated for Vostok 6, and male cosmonaut Valery Bykovsky replacing Ponomaryova to fly launch earlier in Vostok 5 and stay in orbit longer. This resulted in Bykovsky enduring a rushed training regimen, and the flights being delayed by a couple of months.

**Figure 3.5. Tereshkova shortly before her historic flight.**

On June 14, 1963, Bykovsky launched aboard Vostok 5 with Tereshkova in attendance. Her emotions were mixed—she was proud to be a part of something so wonderful and noble, and also afraid that she might not live up to the expectations of her instructors, the Soviet space program, and the USSR. But Gagarin, ever her kind-hearted mentor, whispered to her, "I understand you. It's hard to be the first."[7] He would know, and with that, she felt ready as she watched Bykovsky ascend into orbit.

Two days later, on June 16, Tereshkova and her backup, Irina Solovyova (who had replaced Ponomaryova due to the latter's unacceptable comments about smoking and other controversial ideas), were transported to the launchpad. Before them was the Vostok spacecraft sitting atop the steaming R-7 booster in a cone-shaped launch shroud. Reportedly, Tereshkova did something that would definitely not have been a hit with the Western press by urinating—standing up—on the tire of the transport bus. This was a ceremony performed by the cosmonauts, and she would be "one of the guys" no matter how messy it might get. She reclosed her suit and ascended the fifteen-story launch gantry, where she prepared to enter the spacecraft. After being helped into the capsule, her life support and communications hardware was checked, and then she was sealed inside. Her call sign, as bestowed by Gagarin, was *Chaika*.

Two hours later, Vostok 6 roared off the launchpad, and within minutes Valentina was safely in orbit.

Words born of her rural roots sprang into her mind. "I hear the roar that reminds me of the sound of thunder," she later recalled, "The rocket is shaking like a thin tree in the wind."[8]

Soon after she settled in for the mostly automated flight, she fulfilled her civic duty when she broadcast the required greetings to her fellow Soviet citizens—and a listening world—below.

"Warm greetings from space to the glorious Leninist Young Communist League which reared me," she said. "Everything that is good in me I owe to our Communist party and the Young Communist League."[9] The politicos on the ground were well pleased. Less indoctrinated souls probably rolled their eyes.

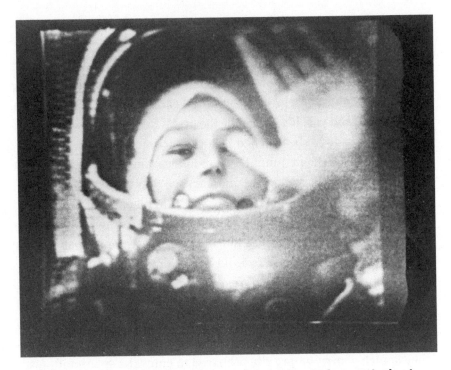

**Figure 3.6. Tereshkova waves to the camera during a televised message from orbit.**

She commented on what she saw, as she had been told to do: "I am *Chaika*. I see the horizon. There is a blue stripe. This is the Earth—how beautiful it is! Everything is going well."[10]

The primary mission was scheduled for one day in space, with the possibility of extending it by two more days if all went as planned. It was touch-and-go for a bit—Tereshkova soon got sick to her stomach and vomited inside the capsule, never a pleasant experience in zero-g. But some male cosmonauts had done the same, so no harm, no foul. She continued on, blaming the brief sickness on the taste of the space food she had eaten. It was better than admitting weakness.

There have been mutterings over the years that she panicked in orbit and begged to come back to Earth early on, but there are no credible reports to substantiate this. Whether these rumors were based on supposition or due to efforts by chauvinistic minds to discredit her, by all reputable accounts, other than the brief nausea she did well during the flight. As the official Soviet space history, published in 1973, put it, she had performed "adequately."

In its first orbit, her spacecraft passed within three miles of Vostok 5. She chatted with Bykovsky while they were within range for radio communication, and a nation cheered.

But there was one glitch, not of her making, that was not revealed to the public for many decades. Within hours of reaching orbit, she realized that the programmers of her primitive computing system had made a serious, and potentially disastrous, error. The retrofire system that would cause her to reenter three days hence had been programmed wrong—it was set to boost her up instead of down. If that occurred, she would have been sent into a higher orbit—which would have been fatal. Flight controllers on the ground scrambled to fix the error and send up new programming.

As an indication of her generosity of spirit, she quietly asked that the offending programmer not be punished for his error, and in exchange she would keep the incident secret, which she did for over thirty years. She never spoke of it publicly until after the technician had revealed the truth himself.

During the second day of the flight, Tereshkova uncharacteristically missed a major step in her checklist. She was supposed to orient the Vostok using manual controls, which was a test to assure that she would be capable of overriding the automatic control system should it fail to properly prepare the spacecraft for reentry. The flight controllers were quite concerned that she had missed this critical step, and they instructed her to try it on the next, and final, day of the flight.

At the conclusion of the third day in orbit, the Vostok's automated system aligned itself for reentry and fired the braking rockets—she was headed home; no manual control was needed after all. She was in a good deal of discomfort—the pressure suits and arrangements inside the capsule had been designed for men, and much of it was not suited to her smaller stature. But it would all be over soon—all she had to endure was a few minutes of high-g loads and she'd be home.

After the flaming fury of reentry, a parachute popped out of the Vostok, and once the spacecraft was stabilized Tereshkova was ejected from the capsule while it was about four miles in altitude, just as Gagarin and the other cosmonauts who flew that spacecraft had been. She glided to the ground as she had hundreds of times, swinging from her parachute as the Vostok capsule slammed onto the tundra under its own chute. They both came down a few hundred miles southwest of Novosibirsk in rural eastern Russia. Tereshkova came to earth about a thousand feet from the capsule and was soon met by a group of farmers. She had to hitch a ride to the nearest payphone to call the Kremlin, and she spoke briefly to her champion Nikita Khrushchev. The flight was portrayed as a smashing success.

Valery Bykovsky, Valentina's companion in space aboard Vostok 5, reentered the atmosphere just hours later, having accumulated 119 hours in orbit, smashing the US record of 34 hours, 9 minutes, held by Gordon Cooper in his Mercury flight.

Within hours the Soviet-controlled media had trumpeted the accomplishment all over the USSR and across the world. The Soviet Union had flown a solo woman in space! Khrushchev was quoted as saying, "Bourgeois society always emphasizes that woman is the weaker sex. That is not so. Our Russian woman showed the American astronauts a thing or two. Her mission was longer than that of all the Americans put together."[11] He was at least partially accurate—her flight had exceeded that of the longest Mercury mission to date, having lasted almost 71 hours.

Tereshkova's Vostok 6 spacecraft was the last Vostok built, and the launch of her mission signaled the end of the Vostok series. From then on, cosmonauts would fly in twos and threes in the updated Voskhod capsule. The later Soyuz craft would carry solo cosmonauts only twice in early development, and would then carry crews of two and three across its long history. (It continues flying today, delivering crews to the International Space Station.)

Upon her return to Moscow, Tereshkova was plunged into a world of international politics, and her role as a cosmonaut became largely representational. She attended the Soviet-sponsored Women's International League for Peace and Freedom as their premiere guest. She then embarked on a grueling goodwill (and propaganda) tour, part of it with Gagarin, across the Soviet Union and to a number of politically aligned countries. Tereshkova was made a member of the World Peace Council in 1966, a Soviet group dedicated to disarmament. In that same year, she was also inducted into the Supreme Soviet of the Soviet Union, the top legislative body of the USSR, a post she held until 1974. From 1969 to 1991 she was a member of the Central Committee of the Communist Party as well. In all these roles she promoted the achievements of Soviet women and advocated communist ideology from her perspective as a Hero of the Soviet Union, the highest honor that could be bestowed upon a Soviet citizen. She was also awarded the Order of Lenin and a number of other medals and honors.

Figure 3.7. Tereshkova was a hit on the public relations circuit after her flight. She is seen here at an international women's conference. (Courtesy of Wikimedia Creative Commons, RIA Novosti archive, image #726670 / Yuryi Abramochkin / CC BY-SA 3.0.)

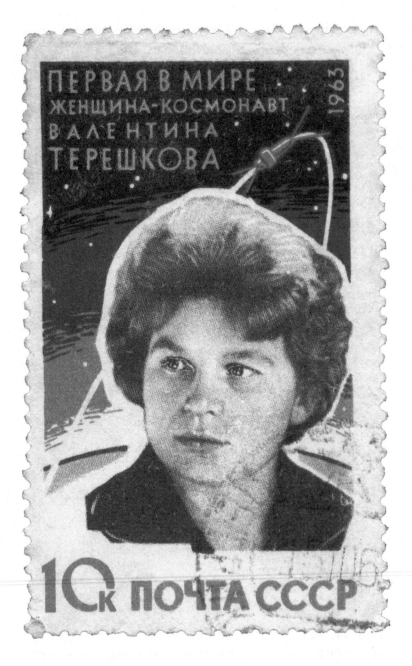

**Figure 3.8. Tereshkova was immortalized further in this 1963 stamp.**

Tereshkova also continued her higher education, studying at a Soviet air force academy and graduating as a cosmonaut engineer, and later, in 1977, completing a doctorate in engineering. She was a Communist woman who had exceeded not just her own dreams, but those of many fellow female citizens. But despite the achievements and accolades, she would never fly in space again. The Soviet Union had made its point and scored its first of sending a woman into space, and for the next twenty years that was that.

She performed her role in Soviet society with acumen and enthusiasm, but it was largely ceremonial and political in nature. But inwardly, she pined for another space mission. The Soviets, however, saw no reason to fly a woman again for many years, and if they did, it would not be Tereshkova. While she did perform adequately during the flight of Vostok 6, records released after the fall of the Soviet Union indicate that she was observed as having been visibly fatigued for much of the mission, had missed a number of objectives, and that she slept more than was planned. It appeared that there might be some benefit of sending highly trained test pilots into space after all, though the matter of gender was still an open question. For his part, Chief Designer Korolev wrote that in hindsight at least two of the other female candidates had in fact been better prepared for the flight, but that neither of them could compare with Tereshkova in the ability to influence crowds and arouse sympathy with the common folk. Since this had in fact been the true essence of the mission—to show the world that this plain-spoken farm girl could conquer the heavens at least as well as the astronauts of the Imperialist West—in that light she had been very successful.

Regarding the lack of future flight assignments, despite many indications that the government simply was not interested in sending more women into space, Tereshkova maintains that she was taken off the rolls of active cosmonauts due to Gagarin's accidental death in 1968. She was told that she was too valuable to the state to risk in another spaceflight. She was certainly not allowed to skydive any longer.

When the Soviet Union collapsed in 1991, Tereshkova's official titles evaporated but her prestige did not. She remains widely regarded as a hero to Russian women, and she is still seen as the approximate equal of Yuri Gagarin and Alexy Leonov (the first spacewalker) in that society. She was later returned to a government role in the Russian State Duma, and continues to serve in that body.

**Figure 3.9. Tereshkova with the "chief designer" Sergei Korolev. She was not his first choice for the flight, but he respected her appeal to the public.**

After her single spaceflight, Valentina Tereshkova married fellow cosmonaut Andriyan Nikolayev, with whom she had one daughter. This marriage was apparently one born more of political expedience than love, and she was given away at the wedding by a beaming Khrushchev who had reportedly encouraged the union. Nikolayev was the only bachelor cosmonaut of the early group, had flown on Vostok 3, and had been assigned to coach the female cosmonaut candidates. Whether or not there was a genuine romance between the two is unknown—both were tight-lipped about the courtship. But after Khrushchev was deposed in 1964, the couple's relationship soured, and while they remained married they did not live together afterward.

They divorced in 1977, and Tereshkova remarried in 1979, this time to a Russian orthopedist named Yuliy Shaposhnikov, who died in 1999.

**Figure 3.10. Tereshkova with her husband, Andriyan Nikolayev, and their daughter, Elena. (Courtesy of RKK Energia.)**

Whatever state intervention there might have been in her personal life, and despite the lack of any flight opportunities after the flight of Vostok 6, Tereshkova remained a staunch Communist until the end of the Soviet Union, and she continues to generally "stick to the script" about her involvement in the Soviet space program.

As far as her comments on equal treatment and the unstinting support of her fellow cosmonauts, the record begs to differ. While Tereshkova continues to assert that the female cosmonauts were treated as equals, Alexey Leonov, who performed the first spacewalk during the Voshod 2 flight in 1965, was quoted as saying in 1975, "When we analyzed the results of her flight afterward, we discovered that for women, flying in space is a hard job and that they can do other things down here. . . . After training, she will be twenty-eight or twenty-nine, and if she is a good woman she will have a family by then."[12] Other male cosmonauts were equally dismissive, saying that women should be in space as scientists and technicians, and "of course, as stewardesses."[13]

It seems that despite Tereshkova's solid performance in space, what would come to be known as the glass ceiling was going to be just as hard to shatter in a powerful rocket as it has been everywhere else. It is worth noting that two of the four women cosmonauts were briefly considered for a flight aboard Voskhod, the two-seat spacecraft that succeeded Vostok, but with Korolev's death in 1966, this plan appears to have been scrapped. The female cosmonaut program was officially disbanded in 1969.

Tereshkova has continued to "fly the flag" for Russia, despite openly pining for the days of the Soviet Union that she claims had treated her well overall. Her daughter married, became a surgeon, and gave birth to a son. Tereshkova continues to live as a single widowed women, working in service to her country.

No other Russian woman would fly in space until 1982, when cosmonaut Svetlana Savitskaya orbited in a Soyuz capsule just under a year before Sally Ride, America's first woman in space, did so aboard the space shuttle. Savitskaya later became the first woman to conduct a spacewalk. These represented some of the final Soviet "firsts" in space, worthy bookends to Valentina Tereshkova's groundbreaking flight.

In 2013, at the age of seventy-six, Tereshkova admitted that her dreams of spaceflight have never abated. It is her dream to go to Mars, she has said at a number of public functions. When it was pointed out that it could be a one-way trip, given the limits of current technology, she said, "I am ready."[14]

Her comments from an autobiography published in 2015 may best summarize her longing for one more spaceflight: "Those who have already been in space, yearn with all their heart and soul to hasten there again and again."[15]

**Figure 3.11.  Tereshkova in Moscow in 2017. (Courtesy of Wikimedia Creative Commons, source: Kremlin.ru, licensed under CC BY 4.0.)**

# CHAPTER 4

# GENE KRANZ: STARS AND STRIPES FOREVER

**E**ugene Francis "Gene" Kranz never said, "Failure is not an option." That was an invention of the screenwriter on Ron Howard's excellent film *Apollo 13*. But it is something that he *could* have said, and may well have thought, because that is who Gene Kranz was, and still is. What he did say, in an address to the stunned controllers assembled before him in a somber meeting room just days after the Apollo 1 fire, a major accident that took the lives of the first three astronauts slated to fly the spacecraft, was even more powerful. Kranz said, in part, this: "From this day forward, Flight Control will be known by two words: 'Tough' and 'Competent.' . . . We will never be found short in our knowledge and in our skills. Mission Control will be perfect."[1]

Perhaps what Ed Harris, who played Kranz in the movie, should have said was, "Imperfection is not an option." Kranz's goal was indeed to make Mission Control, and the flights they oversaw, as close to perfect as was humanly possible.

They got pretty damn close.

Kranz was born amidst the Great Depression, on August 17, 1933, in Toledo, Ohio, and grew up on his family's farm. His childhood was not an easy one; his father died when he was seven, and the family struggled to get by, despite the slowly improving economic conditions in the country. Kranz and his two older sisters, Louise and Helen, worked hard to help their mother get by. The family took in borders to help make ends meet, and many of these tenants were in military service. The ethics and patriotism of these young men seems to have appealed to Kranz and contributed to his own strong sense of duty and love of country—themes that drove him throughout his career.

During the years of World War Two, Kranz delivered newspapers, and used the opportunity to keep abreast of the goings-on in Europe and the Pacific, charting out major battles and developments with articles pinned to his bedroom wall. Like John Glenn, he also developed an early interest in flying, building models of airplanes from wood strips, tissue paper, and glue. He later flew model rockets of his own design, and devoured the writings of early thinkers in spaceflight such as Willy Ley and Wernher von Braun—some of whom would later be his colleagues at NASA.

Kranz was the product of a Catholic high school education—a strict regimen in those years. His early interest in spaceflight—which would not actually occur for almost a decade—is reflected in his choice of topic for his high school thesis, entitled "The Design and Possibilities of the Interplanetary Rocket," which he wrote in 1950. The sixteen-page paper was illustrated with beautifully rendered drawings by Kranz, showing details of the German rockets from World War Two. (It must have been quite a moment when Kranz met von Braun, the creator of those rockets, a number of years later.) The paper rated a comment of "Very Good" from his advisor, with a score of 98 percent, and in it, he prophesized that humans would land on the moon by 1960. "An examination of the current technical and industrial development demonstrates the high probability that the Moon will shortly be conquered by man. The base will probably be established in five years and completed in ten," he opined.[2] "I missed it by a decade," he later said with a chuckle in a 2010 interview.[3]

Kranz went on to complete a degree in aeronautical engineering at St. Louis University, financed by his own savings, a small scholarship, and his family's sale of his father's prized stamp collection and other possessions. The highlight of his college experience was Kranz' participation in their basic flight training program, flying outdated Stearman biplane trainers. Regardless of the aircrafts' vintage, he was thrilled to be in the air. The next stop would be jet fighters.

In 1954, a year after he graduated from college, Kranz was commissioned as a second lieutenant in the air force. While waiting for his flight assignment, he landed a job at McDonnell Aircraft, working as a flight test analyst, compiling results from testing of some of the most advanced aircraft of the day. The experience of distilling notes and readouts into understandable and useful data from a test flight would come in handy once he transitioned into NASA.

He soon got his flight assignment with the air force, and relocated from one base to another to complete various aspects of his training. These postings ranged from Texas to Georgia, back to Texas, and finally to Nellis Air Force Base in Nevada, where he trained in the newest jet fighters.

After a new posting in South Carolina, he married Marta Cadena, the daughter of Mexican immigrants, whom he had met earlier while in Texas. But within months Kranz was dispatched to his duty posting in Korea, just two years after the end of that conflict. There he flew fighter jets along the border between North and South Korea, part of the international effort to maintain security in the tense region. It was the beginning of the many separations they would endure as a couple during his career, but these periods never seemed to interfere with their love for one another, the underpinning of a relationship that continues to this day.

Upon the completion of his military service, Kranz returned to McDonnell, working on missile programs in St. Louis, then in flight test in New Mexico. Kranz loved flying, and the work appealed to his interest in engineering. During this time, he saw an advertisement in *Aviation Weekly*, an aerospace publication, for a new government agency called NASA. He applied immediately, and by 1960 he was working with the newly created Space Task Group at NASA's Langley Research Center. He was now twenty-seven years old.

Kranz was assigned to Christopher Kraft, a central figure in the planning of America's entry into the space race, and later in the effort to send astronauts to the moon. Kraft immediately put Kranz to work writing flight procedures manuals for the upcoming Mercury missions—Kranz literally "wrote the book" guiding US spaceflight in the upcoming decade. This was a perfect introduction for the young engineer to the world of being a Flight Director, a post he would assume in just a few years.

**Figure 4.1. Gene Kranz poses with his F-86 fighter jet during his military service. (Courtesy of the United States Air Force.)**

Putting humans into space was a new endeavor, and as Kranz began his work with NASA, everything was new and untried. "The man in space program was simple in concept, difficult in execution," he recalled in his autobiography, *Failure Is Not an Option*. "Every mission was a first, a new chapter in the book. Many, if not most, of the components in both rockets and capsules had to be invented and handmade as we went along, adapting what we could from existing aviation and rocket engine technology."[4] He described leaving a world in which jet airplanes moved at five miles per minute and entering as era of unparalleled speed. "In this new, virtually uncharted world we would be moving at five miles per second," he said.[5]

This was NASA's first manned spaceflight effort. The Mercury flights would be controlled from Cape Canaveral in Florida—this was before Mission Control was located in Houston at what would become the Johnson Space Center, and before there was a Kennedy Space Center in Florida. Mercury Control was located at the Cape Canaveral Air Force Station, not far from the launchpad from which the Mercury flights would depart. It was cutting edge for the time, but by today's standards very basic. They did not even have computers there yet—the huge mainframes needed to control the flight were located at NASA's Goddard Space Flight Center, in Maryland. Data processed by these massive machines was converted into instructions that were then sent to Mercury Control.

As Kranz got to know his way around the facility, and around early spaceflight in general, Mercury was having teething pains. The program was, by 1960, already about a year behind schedule. The rockets that would be used had been identified, but they had serious issues. The Redstone booster was reliable but only powerful enough for suborbital tests of the Mercury spacecraft, including two manned sub-orbital tests with astronauts to come in the following year. The second rocket, the Atlas intercontinental ballistic missile, was a far larger and more powerful rocket—and that much more complex and troublesome. As discussed in chapter 2, it was made with a thin stainless steel skin, was delicate to handle, and had a nasty habit of exploding after launch—as many as half of them failed in flight.

Things were tense around the Cape.

By November, the first test flight of a complete Mercury spacecraft atop a Redstone was ready to go. There would be no astronaut, nor a monkey, riding inside—this was an unmanned test of the machine. The simple one-stage Redstone rocket, with just a bit more thrust than the X-15 flying out at Edwards Air Force Base in California, sat out on the pad, and all systems had checked out. Kranz had only been on the job for just over a month, but he felt optimistic about the flight. At T-minus-zero the ignition signal was sent to the rocket, and as the assembled technicians watched in tense anticipation, the booster belched smoke and flame— and then just sat there, doing nothing more.

Frantically spoken German could be heard in the control room—many of the technicians were from von Braun's original team brought over from Germany after the war—and Kraft had to demand some answers in English. It soon became clear to everyone that the rocket engine had ignited, the booster had left the pad, and then the whole thing suddenly lost power, settling right back onto the launchpad. It came to rest just as it had departed, bolt upright. The total altitude for the flight of Mercury-Redstone 1 (MR-1) had been about four inches. To make things more embarrassing, as the automatic flight timer onboard the rocket continued to count-off the seconds, and the escape tower, a small emergency rocket atop the Mercury capsule, had fired right on cue, blasting off into a far more successful flight of its own. It crashed about 1,200 feet from the launchpad, after reaching a peak altitude of 4,000 feet. In one final indignity, the parachute popped out of the top of the capsule a few seconds later, fluttered upward for a moment, then draped itself around the top of the still erect booster.

It would have been laughable had the stakes not been so high. There America's pride-and-joy sat—a fully fueled bomb, still armed and ready to blow for all anyone knew, and communicating nothing to the control center—there were no signals coming from the Redstone at all. The range-safety devices, small bombs located at various parts of the rocket, were still armed and could go at any minute, creating a massive explosion on the pad, and could not be shut off.

It soon became clear that the rocket had lifted just far enough to pull free of the cable that transmitted information between the rocket and the control center, and at that instant the onboard computer had decided to shut down the engines. The cable was now dangling below the rocket, unplugged and useless.

The German technicians discussed a number of ways to try to defuse the situation. One idea was to send a technician out to the pad to crawl under the rocket and plug the umbilical wires back in so that the controllers could perform an orderly shutdown. However, the very act of plugging the cable back in could cause a short circuit, and the rocket could blow, killing the technician. Another idea floated was to drive a crane truck out to the pad from which an equally brave technician could cut the cords that still connected the still fluttering parachute to the Mercury capsule atop the rocket—at least with the parachute removed, there would be no chance of an errant breeze inflating it and pulling the rocket over, which would also likely result in an explosion.

But the most interesting proposal from the German rocket team—ever pragmatic—was for someone to get a hunting rifle and shoot enough holes into the fuel tank to allow the fuel to drain free, at which point they could ostensibly go out and disassemble the crippled booster, holes and all—if it didn't explode from the gunshots.

By now Kraft had endured enough. He was furious. On the spot, he informed the assembled crew of the first rule of flight control—if you don't know what to do, *don't do anything*. This dictum was burned into Kranz's mind on that day, and would be repeated countless times before the space race ended with the first landing on the moon.

They waited overnight, and by morning it was a machine that could be safely dealt with. The rocket's batteries had drained to nothing, and the technicians could now disarm it and drain the kerosene fuel without the help of bullet holes (the liquid oxygen oxidizer had boiled off overnight as it warmed). The rocket was in acceptable condition, as was the capsule. Everything was carefully examined, and a month later the test flight was completed successfully. But the press had enjoyed a small field day with the news of the first failure—just one in a series of black eyes for America's fledgling man-in-space project.

But things improved around the Cape after the successful test, and as the Mercury program gathered steam, Kranz's duties became more defined. Besides writing the mission procedures and rules, he was to assure that everything possible that *could* be done before a flight *had* been done to assure success. This included preparing all the procedures and messages that would have to go back and forth between the control center and all the tracking stations around the world—NASA did not yet have its global network of tracking antennas in place, so a handful of fixed radio dishes around the globe, along with seagoing naval vessels equipped for radio communications, had to be cobbled together into one big, relatively seamless system, capable of tracking a spacecraft as it circled in orbit. It was a heady assignment for Kranz, still in his twenties; one filled with stress and uncertainty. He loved it.

It was now 1961, and the first flight of a manned Mercury spacecraft was scheduled for April. So was the newest addition to Gene Kranz's growing family; Marta was pregnant with their third child—there would eventually be six children—and due to deliver soon. Then, before astronaut Alan Shepard could complete his historic flight—which would have made him the first human in space—the Soviet Union launched Yuri Gagarin into his one-orbit mission on April 12, and efforts to get an American into space—even just a suborbital flight as Shepard's would be—were redoubled.

These first flights of the Mercury program were all run out of NASA's ramshackle facilities at Cape Canaveral. It would not be until the early Gemini flights that what most of us think of as "Mission Control" would be constructed in Houston at the Manned Spacecraft Center (or MSC), later the Johnson Space Center. Nonetheless, while working in the primitive conditions at the Cape, the core elements of what Mission Control would be, and how it would work, were taking shape under Kraft and, increasingly, Kranz. Central to those efforts was flight simulations—almost an early form of virtual reality, Kranz would later say—that were so close to the real thing that by the time you flew a mission, you would feel as if you had done it ten or a hundred times before—just another simulation, or in the vernacular, a "sim."[6]

Early on, one group of flight controllers would create and run a simulated mission for another group of flight controllers. There was a person put in charge, called the Simulation Supervisor, or "SimSup," as they quickly became known, to manage the process. At the time of the Mercury program, these simulations played out almost like a radio drama. The people running the simulation would pretend to be people scattered throughout the simulated mission—offshore receiving dish operators, naval recovery ship communications, you name it—and would also provide simulated telemetry from the simulated spacecraft, simulated errors in the ground equipment, and more. It was important, they felt, to run the flight controllers and technicians who were being tested through every possible, conceivable scenario; all the better to prepare them for potential complications in a real mission. As Kranz put it, "The SimSup's objective was to test the judgment of each individual and the competence of the total mission team. How quickly would they recognize and solve problems? How well did the mission rules and the procedures used in the various facilities and the network function in real time? Were we ready?"[7]

An astronaut sitting in a spacecraft simulator was on the receiving end an identical set of problems and communications as managed by the SimSup and his team. It was an incredibly complex process.

While not terribly convincing early on, the simulations improved as the Mercury flights neared their conclusion. As they became increasingly demanding, Kranz would psych himself up before each one with an early-morning round of military marches from John Phillips Sousa, his favorite composer. He'd listen to Sousa during his drive into Canaveral from the shabby beach towns to the south where the controllers stayed. Later, he'd play the brassy marches in his office as he drank his first coffee of the morning. It was corny, cliché, and also Gene Kranz to the core. Like John Glenn, he was not just a patriot (as were many in the space program), but an idealist as well. Sousa lit his fuse for the day to come.

On May 5, 1961, Mercury finally spread its wings. Alan Shepard's quick fifteen-minute suborbital flight was successful after weeks of technical problems and launch scrubs. The second suborbital Mercury flight, piloted by Gus Grissom, followed barely two months later. But it would be seven long months before John Glenn's much more complex three-orbit mission on February 20, 1962, and Kranz and his team were busy as hell, preparing to transition from fifteen-minute lobs to an extended orbital spaceflight lasting hours.

During the short lull between Shepard's and Grissom's flights, there was another pivotal event that changed everything. President Kennedy delivered his first "moon speech" just three weeks after Shepard's flight, committing the United States and NASA to landing a human on the moon before the end of 1969. While those working in manned spaceflight were already running full-bore, this upped the stakes significantly. A program that had merely been desperate to catch up with the Soviet Union to put a man in space—less than 100 miles overhead—was now aiming for the *moon*—240,000 miles away. It seemed all at once magnificent, and yet vaguely impossible. As Kranz later put it, "It seemed, at that moment, like a pipe dream. I thought, 'Well, let them get on with their great plans; I'm gonna get a man into orbit first.'"[8] A short while later, Kranz saw Kennedy in person during a visit to the Cape, and his impression shifted. Of Kennedy, he said, "His energy and charisma were electrifying; he made believers out of all of us, even the most skeptical. Our hopes had been renewed; maybe Kennedy really understood the towering odds we faced—and were willing to overcome."

The gauntlet had been thrown down, and while the teams at Mercury Control continued their efforts, there was a new energy to the program: America was going to the moon.

After Grissom's flight and prior to Glenn's, Kranz had experienced a realization: he had not been pulling his weight, in his own estimation. He knew that they were all entering a new phase of flight, and that at this point, NASA was still operating much as the air force had for high-speed aircraft testing. With Glenn soon to go into extended orbits of the Earth, this was simply not good enough, and things would have to be different. He resolved to up his game.

Kranz also felt a strong allegiance to Kraft and became determined to become his trusted right-hand man. He threw himself into learning every system and subsystem of the Mercury spacecraft and the troubled Atlas booster, and every detail of the proper operation of these exotic new machines, in order to predict and understand even the least likely problems that could occur during the upcoming flights. In essence, he was trying to outthink the SimSups, an endeavor that would serve him well through the end of the Apollo moon landings, and into Skylab and the early days of the shuttle program.

He later said of this shift in his thinking, "An engineer can explain how a system should work (in theory) but an operator has to know what the engineer knows and then has to know how the systems tie together to get the mission accomplished. If the systems break down the operator must make rapid decisions on fixing or working around the problem to keep the mission moving."[9] It was a quick self-imposed education in moving from theory to reality; from design and test to spaceflight operations. As all this was underway within his mind, Marta delivered another daughter. Then it was back off to the Cape, with his wife holding down the fort as she had been for years already. He admired her strength, and counted on it to support his own.

By the time John Glenn launched on his Mercury flight, Kranz was immersed in an entirely new ball game in flight control. He had been made responsible for tracking and communications for this mission, and his organization—the teams, their hardware, and the procedures required to maintain reliable tracking and communications with Glenn's spacecraft as it circled the Earth—were working well. Being the first manned orbital mission for NASA, maintaining contact was critical. Previous orbital flights with primates had helped to refine this globe-girdling system of communications, but the need to maintain constant contact with Glenn overrode any previous efforts. Beyond the fact that there was a human in this spacecraft, and that Mercury Control would need to communicate with him frequently just to assure a smooth flight, the doctors were also worried that an astronaut in orbit might panic at some point or suffer some other space-related malady. Kranz thought this idea was foolish, but he made sure that everything came off like clockwork regardless. It was what he had committed to do.

The teams were spread out across continents and oceans. The CAPCOMS on the ships at sea—usually older vessels that had been modified for this role—had to contend with stormy weather and weeks at sea. Other stations, such as the one in Nigeria, had to contend with restless political conditions. Communication with these far-flung outposts was never great in those days, and keeping it all together for even a short flight such as this one was challenging.

For example, while the remote tracking stations were able to com-municate with the Mercury capsule as it passed overhead, they were often not able to maintain solid voice links with Mercury Control over the telephone lines. In the event that this system failed, the remote stations would relay messages to the Cape via teletype—essentially an electric typewriter/printer that would receive basic coded impulses through a cable and print out the message on paper. A small team would then grab the printouts and cut them into strips with scissors, taping them up on the wall next to a mission timeline so that everyone knew what was going on. It seems like horse-and-buggy technology today, but that was what they had available to them, no matter how much they spent.

Then came the landing bag emergency on Glenn's flight. In the middle of the mission, an indicator lit up on a control console, indicating that the landing bag on the capsule appeared to have deployed early (see chapter 2). If that was true, once the retrorocket pack was released the heat shield could come loose during reentry, with possibly fatal results. The Mission Control team went into a frenzy first locating the appropriate engineers and technicians to help them determine what this signal might mean—was it a real problem? Or was it a bad sensor on the spacecraft? Perhaps it was an error in the ground equipment? At that moment, Kranz realized there was a critical omission in their mission planning. "Precious time was lost trying to track down engineers at the blockhouse and Hangar S," Kranz later recalled.[10] "Kraft's controllers had no provisions for emergency access to the total design, manufacturing, and assembly team." Kranz made a note to himself—make sure you know where to find *everyone* involved with *every system* during a flight, in case you need to get information from them quickly. It was just one of many, many learning moments during the early days of flight control. Lessons such as this one would serve them all well during urgent situations in the Gemini and Apollo programs soon to come.

The landing bag deployment indication had now come in from a second tracking station—it was not an instrumentation error on the ground after all—the signal was coming from the spacecraft. While the issue ultimately proved to be a faulty sensor switch, the controllers were deeply concerned throughout the rest of the mission. This was exacerbated by the fact that queries sent to the remote tracking stations via teletype could take up to fifteen minutes to get answered. Kranz realized they had a long way to go before they could be able to control complex missions such as Gemini and Apollo in real-time.

There would be three more Mercury flights before moving on to the Gemini program, but Kranz later recalled one more major realization that came to him during Glenn's mission: "A controller lives or dies based on the information he has at his console. If you lack what you need at liftoff, there is little hope that you will get new information that you would trust during a mission. This realization was the most profound impression branded on me from the Glenn mission."[11]

By the end of the Mercury program the Mercury Control team numbered about 705 people. When they completed moving the control center to the new facility in Houston, that number was nearly 6,000. America's space program was expanding and evolving quickly, and the Gemini program was finally underway—the lunar landing goal drove everything, and delays were unacceptable.

In this period of time, people advanced rapidly in the ranks of Mission Control—at the pace the space race was going, a couple of years of experience warranted the kind of promotion that might have taken a decade or more in corporate life. Kraft became the chief of the Flight Operations Division, and Kranz was soon promoted to head Flight Control Operations. He had certainly earned his stripes, and the pressure was on to keep Gemini moving.

As they settled their families into freshly built homes near the Manned Spaceflight Center, and themselves into the new Mission Control center, it was clear that the flight control personnel had become a true team. Some of the original members had opted to leave the cadre—they either did not want to move their families to Texas, or just felt that the work was too stressful. Of the remainder, Kranz would say, "Outside of wartime, I do not believe that young people had ever been given responsibilities so heavy or historic. We were in jobs that appealed to the adventurer, dreamer, and Foreign Legionnaire in each of us."[12] In his mind, the best of the military brotherhood was blended with cutting-edge flight in this dream job.

The first manned Gemini flight, Gemini 3, was to be the final mission controlled from the Cape. Gemini 4, which launched on June 3, 1965, was the inauguration for the new Mission Control Center in Houston. While NASA was still using a somewhat ad hoc arrangement of tracking stations to control and communicate with the Gemini spacecraft, communications on the ground were now digital, with much higher-speed and more reliable ability to transmit more information. Additionally, due to the experience gleaned from Mercury, the tracking sites for Gemini had been reduced from thirteen to just six. It was altogether a more refined, compact operation, with much better capacity.

As the Mission Control team settled in, Kranz had another surprise in store for his people—a Gemini cockpit trainer. While just a rough plywood-and-cardboard replica, it would give them a feel for what the astronauts were doing and experiencing, and it was a brilliant addition to their regimen. He went so far as to have them practicing inside the mockup while blindfolded until they knew the cockpit cold.

By the time Gemini was underway, Kranz, now in his early thirties, had been designated as a flight director, along with Glynn Lunney and John Hodge. The Gemini missions would up the orbital ante from many hours to multiple days, then weeks, so extra shifts were going to be required at the control center.

The flights of Gemini 2, the final unmanned test of the system, and Gemini 3, the first flight with a crew, both controlled from the old Mercury Control Center at the Cape, had gone well. The unmanned test flight went off without a hitch, and Gemini 3 followed, with Grissom and rookie astronaut John Young in the cockpit for a three-orbit shakedown. The only real snag was during the flight of Gemini 2—the control room lost power right as the Titan rocket launched, due to the press having overloaded the electrical circuits with their power-hungry movie lights. From that point onward, Kranz dictated that the circuit breakers in Mission Control be locked in place during a mission—he preferred the potential overload, and even fire if necessary, to a dangerous blackout in the control center that could endanger the mission. Another lesson learned.

The flight of Gemini 4 in early June 1965 was Kranz's first as a flight director, and the was the first mission run out of the new Mission Control Center in Houston. Jim McDivitt was the mission commander, and the flight had far more complex objectives than the Mercury missions, including the first American spacewalk, to be performed by astronaut Ed White.

**Figure 4.2. Gene Kranz during the flight of Gemini 4. Chris Kraft is to the right. (Courtesy of NASA.)**

This flight also went by the numbers, and White's spacewalk was a sensation in the Western press, with spectacular photos taken by McDivitt from the cockpit. Gemini 4 also hosted another first: Kranz's wife made him a bespoke white vest for the mission, which he donned with some flourish as his first shift began. Each of the three flight directors had chosen a color to represent their team, and Kranz's color was white ... so, then, the fancy white vest. Here began a tradition of

Kranz unveiling a new vest for each mission, each one resplendent in its own fashion.

**Figure 4.3. Gene Kranz, looking dapper in one of his wife's custom-made vests for Gemini 4. (Courtesy of NASA.)**

Gemini 5 launched on August 21, 1965, with Mercury astronaut Gordon Cooper in the command seat, and rookie Pete Conrad to his right in the cramped capsule. After a fine launch, the troubles began almost immediately—a fuel cell, a new technology designed to augment the batteries that had powered earlier spacecraft, was having problems. Fuel cells mix hydrogen and oxygen and run them past a catalyst, with pure drinkable water and electricity as the byproducts. They were a perfect solution for powering spacecraft—and completely experimental in an orbital environment. When the cell started to malfunction, it looked as if the mission would have to be curtailed far earlier than the planned one-week duration.

When Kranz arrived for his shift, the spacecraft was already in orbit with a failing fuel cell. Kranz saw that Kraft was preparing to leave for the usual post-launch press conference, and he asked him how he wanted to proceed. Kraft looked at him and said, curtly, "You're the flight director, it's your shift. Make up your own mind."[13] Kranz was flummoxed as Kraft departed, then realized that his boss was correct. Kranz was the flight director now; it was his show. Within hours, Kranz and his team pulled in the manufacturer of the fuel cell to see what might be done to correct the problem, and before the end of his shift they had worked up a potential solution. Over the next few days, the team got the balky power generator working again, and Gemini 5 limped through the mission, as Kranz put it. The crew was able to stay in space for the full week, breaking the record previously set by the Soviets, and Gemini 5 was another success.

The remaining seven Gemini missions flew over the next fifteen months, with the mission objectives necessary to move on to Apollo being checked off one after the other. Their prime objectives focused on long duration spaceflight, rendezvous with a second Gemini capsule, and rendezvous and docking with a second, unmanned spacecraft called the Agena. There had been a close call during Gemini 8, piloted by Neil Armstrong with David Scott in the second seat, when a thruster stuck and the spacecraft spiraled out of control for a few minutes, but in that case the crew made all the right decisions to save themselves and the mission, coming home early but safely (see chapter 6). There was only one major hurdle to clear before declaring the road to the moon open: the dark art of Extra Vehicular Activity, or EVA.

Since the beginning of the planning process to reach the moon, EVA—or, more informally, spacewalking—had been a part of the skill set deemed necessary for the Apollo missions. It was thought early on that they might have to transfer from one spacecraft to another in open space, or that there might be the need to repair a spacecraft while en route to or from the moon. In either case, astronauts would not only have to be capable of exiting and entering a spacecraft in a hard vacuum, they might also need to be able to move from one craft to another and compete useful work to fix some malfunctioning system. This was viewed as challenging but something that would be quickly overcome. How wrong they were.

Ed White's spacewalk in 1965 had been a breeze. He left the capsule, maneuvered around with a small hand-held thruster, and returned to the cockpit. The next EVA exercise had been planned for Gemini 8, but this was canceled due to the in-flight emergency that caused the crew to return to Earth early. Then EVAs were conducted on Gemini 9, 10, and 11, but with decidedly mixed results—the astronauts working outside the spacecraft tended to become winded and then exhausted, necessitating an early termination of their activities. While each flight achieved incremental gains over the previous one, the goal of smooth transition from one spacecraft to another, and of doing actual work once they arrived, eluded NASA. It was frustrating and concerning, as this final aspect of EVA was considered critical to moving ahead with Apollo. The EVAs were almost more frustrating for Kranz and the other controllers, who knew that mastering them were critical to the future of the manned spaceflight program, but who could only look on in frustration from Mission Control, unable to do much more than monitor the attempts and make suggestions for something they could not even see—there was not yet any live TV from spaceflights.

The first chance to attempt a more complex EVA came on Gemini 9. Gene Cernan would exit the spacecraft, while Tom Stafford manned the controls. Cernan was to move to the back of the Gemini capsule and don a maneuvering backpack built by the air force for testing in free flight (though he would still be tethered to the Gemini capsule via a long cord). But almost immediately upon exiting the capsule, Cernan started to pinwheel in space and struggled to get a grasp on the spacecraft to haul himself to the back section, near the heat shield, where the maneuvering unit was affixed. By the time he got there, his heart rate had peaked at 180 beats per minute, and his helmet's visor was fogged with condensed perspiration. Stafford called the remainder of the EVA off and Cernan returned to his seat inside, completely exhausted.

In Mission Control, Kranz had watched the entire proceedings with increasing apprehension. It was ultimately Stafford's call whether or not to continue the EVA, and Kranz was greatly relieved when the EVA was terminated, thinking, "God, those guys are like icemen, chock-full of guts!"[14]

**Figure 4.4. Gene Kranz at the console for a Gemini test procedure in late 1965. (Courtesy of NASA.)**

At the end of Gemini 9, Kranz moved his efforts to the preparation for the first Apollo flights, scheduled to begin in just a year. The other flight controllers would complete the Gemini program, and Kranz found that he envied them. He did manage to pull a few of the overnight shifts to give the two remaining flight directors a break—they could now have a regular eight-hour shift (nine hours including handover and debriefs) instead of trying to do twelve hours each. But the other flight directors, Glynn Lunney and Cliff Charlesworth, would complete the challenges of the Gemini program, and ultimately master the intricacies of EVA, with Buzz Aldrin turning in a by-the-letters and perfect performance on Gemini 12. Aldrin had experimented with underwater training in a local high school pool, which simulated weightlessness pretty well, and months earlier had strongly stated his opinion that this was the way to proceed. A NASA headquarters review of the issues in Gemini EVAs agreed with his assessment.

When Gemini 12 splashed down on November 15, 1966, the Gemini program was officially over. The first manned Apollo flight was scheduled for just over a year later, and Kranz was already deeply engaged in preparations to oversee the operation this new and vastly more complex spacecraft. He was now the chief of the Flight Control Division, and refining the procedures for running the Apollo missions in their entirety would be his responsibility.

By the end of 1961, after a few years of working with McDonnel Aircraft on Mercury and Gemini, NASA, and therefore Kranz, had a new contractor to deal with, North American Aviation. North American had built some of the most potent fighter aircraft of World War Two, as well as the X-15 rocket plane. Now the company was building the Apollo capsule (called the Command Module or CM) and the unit that powered and propelled it, which was called the Service Module (or SM—the combination was called the CSM). North American had a somewhat different business culture than the other contractors NASA had worked with, however—they were not as welcoming of NASA's oversight, and tended to be more independent. The relationship became challenging—by 1966, Kranz was butting heads with them over the creation of procedures manuals for the Apollo spacecraft. In the following year the partnership would become much more highly charged.

Apollo 1 was scheduled to fly with its first crew in early 1967. Gus Grissom and Ed White were both veteran astronauts (both had flown in Gemini, and Grissom had also flown in Mercury), and the third crew-member, Roger Chaffee, was a rookie. The Apollo capsule they would fly was an early design called the "Block I." The capsule had been redesigned since the Block began construction in order to accommodate docking with the Lunar Module that would carry two astronauts to the moon's surface. But since NASA had an apparently flight-worthy Block I spacecraft ready to go, the decision was made to fly the older design at least once to get experience with the CSM in Earth orbit—no Lunar Module (LM) would be on this first flight.

As the launch date approached, Kranz had a nagging feeling that he and his teams were "behind the aircraft," which is to say that he felt they were not really on top of all the systems and procedures yet. There was not much time left to get things right—1966 was ripening.

Apollo 1, as this first mission later became known (technically, it was called Apollo-Saturn 204 (AS-204), was intended to test the launch operations, spacecraft tracking and control, and of course the Apollo spacecraft hardware and its operations. It would launch on a Saturn IB rocket, a smaller predecessor to the Saturn V moon rocket. The mission was supposed to last up to two weeks if all went well . . . but the crew never had the opportunity to launch.

On January 27, 1967, NASA was conducting a test of the spacecraft, which was already mounted atop the Saturn IB booster in preparation for launch on February 21, with the crew inside the capsule. This was a critical "plugs out" test, in which all the cabling and connections between the spacecraft and the launch tower were unplugged to test the system in flight configuration. It was considered a low-risk test, since the rocket was not fueled. This was primarily a procedural prelaunch shakeout between the spacecraft and the ground controllers and their equipment.

Kranz remembers the day with great sadness. The test had started early in the morning, and Kranz and his team were monitoring the test from Mission Control in Houston. "I had the responsibility for the Mission Control team, the Mission Control Center, communications, the remote site teams, etc.," Kranz later said. "I had checked out all of the communications. I'd talked with the launch team down at the Cape. We'd picked up all the deviations to the procedures and had worked through into the early afternoon time frame." At that point, another flight director, John Hodge, arrived to relieve Kranz. "I had handed over the console responsibilities to John, and he was going to continue the countdown until the time frame when we got very close to the simulated launch, and Kraft would come in and pick up the count."[15]

It was noon, and Kranz was ready to end his shift. "The crew had entered the spacecraft, and everything looked like it was going reasonably well for a change. We had had problems in communications, but nowhere as severe as we'd had in the previous day's testing." Kranz went to his office to wrap up his day, then headed home. He wanted to take his wife out for dinner—something that occurred rarely due to the crushing schedule he and thousands of others as NASA were keeping. "Marta had our third child, so I'd promised her an evening out for a change."

Then at 5:31 Houston time, the controllers monitoring the test heard a frantic message from the crew inside the Apollo capsule.

"Fire!"

"We've got a fire in the cockpit!"

"We've got a bad fire . . . get us out. We're burning up . . ."

In less than thirty seconds the astronauts were dead, asphyxiated by a flash fire that roared through the small capsule.

Kranz was at home preparing for his outing. "We were in the process of dressing to go out, and my next-door neighbor, Jim Hannigan, came over. Actually, we were waiting for the babysitter, and we heard some loud pounding on the door. . . . [He] came in and identified that he had heard over the radio that there was a serious accident in the launch complex and suspected the crew was dead. So this was the first indication I had that we had the Apollo 1 disaster."[16]

Kranz scrambled to his car and headed to the space center:

I arrived out there, and they had secured all of the doors, and there was no way to get in on the phone from the security people up to the Mission

Control floor. I kept circling around the building, and there was a freight elevator back there, and I sort of buffaloed the security guard to get access to the freight elevator and up to the floor where we were conducting the test, and got up there and got into Mission Control.

I've never seen a facility or a group of people, a group of men, so shaken in their entire lives. Kraft was there. He was on the phone down by the flight surgeon talking to the people down at the Cape, I believe Deke Slayton. John Hodge and I had grown up in aircraft flight tests, so we were familiar with the fact that people die in the business that we were conducting, so we had maintained maybe a little bit more poise relative to the others, but the majority of the controllers were kids fresh out of college in their early twenties. Everyone had gone through this agony of listening to this crew over the 16 seconds while they—at first we thought they had burned to death, but actually they suffocated, but it was very fresh, very real, and there were many of the controllers who just couldn't seem to cope with this disaster that had occurred.[17]

Kranz ordered the controllers to secure their consoles, and told them to make detailed notes of everything they could recall. When this was completed, most of them headed over to a local bar called the Singing Wheel, their regular watering hole, and numbed the pain and shock with alcohol. When the owner of the restaurant heard the news, he banished all the other customers from the building—that night would be a private wake for the crew of Apollo 1.

The next morning, we came back out to work again, trying to see if there were any answers, because in that kind of an environment you're trying to find answers, you're trying to find out why, what happened.... We worked through the Sunday time frame, again just sitting in offices almost just paralyzed. We were so stung.

By Monday morning, Kranz had made a decision—he felt that he needed to contextualize the incident for his flight controllers. These were young people—their average age was just twenty-six—who had not faced this kind of loss before. He also wanted to get his department back on track—he had determined that, as a group, they had drifted from their core responsibility.

He collected his people in a meeting room, and waited for the group to quiet down. Other NASA employees, and a number of people from the spacecraft contractor teams also filtered in. The mood was understandably somber. After a steely look around the room, and a few introductory remarks, he laid out the "this is how it's going to be" part of the speech, the most famous in the history of Mission Control, told here as he recalls it—the meeting was not recorded.

> Spaceflight will never tolerate carelessness, incapacity, and neglect. Somewhere, somehow, we screwed up. It could have been in design, build, or test. Whatever it was, we should have caught it. We were too gung-ho about the schedule and we locked out all of the problems we saw each day in our work. Every element of the program was in trouble and so were we. The simulators were not working, Mission Control was behind in virtually every area, and the flight and test procedures changed daily. Nothing we did had any shelf life. Not one of us stood up and said, "Dammit, stop!" I don't know what Thompson's committee [an accident review board chaired by Floyd L. Thompson, the director of NASA's Langley Research Center] will find as the cause, but I know what I find. We are the cause! We were not ready! We did not do our job. We were rolling the dice, hoping that things would come together by launch day, when in our hearts we knew it would take a miracle. We were pushing the schedule and betting that the Cape would slip before we did.
>
> From this day forward, Flight Control will be known by two words: "Tough" and "Competent." *Tough* means we are forever accountable for what we do or what we fail to do. We will never again compromise our responsibilities. Every time we walk into Mission Control we will know what we stand for. *Competent* means we will never take anything for granted. We will never be found short in our knowledge and in our skills. Mission Control will be perfect. When you leave this meeting today you will go to your office and the first thing you will do there is to write "Tough and Competent" on your blackboards. It will *never* be erased. Each day when you enter the

room these words will remind you of the price paid by Grissom, White, and Chaffee. These words are the price of admission to the ranks of Mission Control.[18]

The assembled flight controllers were speechless. While lowering the boom, he had also successfully welded his broken team back into a cohesive whole. Kranz, the man who had long admired Chris Kraft's ability to galvanize people with a few well-chosen words, had just made a speech that went down in the history of Mission Control and NASA, and is studied as a prime example of strong leadership to this day.

NASA initiated a painstaking investigation of the accident, and work on the Command Module, along with preparations for upcoming flights, came to a halt for about eighteen months.

After carefully disassembling the Apollo 1 spacecraft, the suspected cause of the fire was soon found. There had been a wire bundle near the floor of the capsule that was near a compartment door. That door had been opened and closed many, many times, and its sharp edge had scraped against the wires, removing insulation and allowing a short circuit that caused a spark. In normal conditions this would not be a big deal; it might have blown an electrical breaker. But the atmosphere used inside the Command Module was pure oxygen, which is highly flammable. At the five pounds per square inch that the spacecraft would be holding in space, the small flame that would have resulted would probably have been extinguished quickly—the crew had a fire extinguisher for just such emergencies. But because the capsule was at sea level during the test, it had been pumped up to pressures ranging from 14 to 17 psi, at which oxygen is explosive if exposed to an ignition source. To make matters worse, the inside of the spacecraft had strips of Velcro glued in many places—the astronauts used it to secure items that would otherwise float away in zero-g. The Velcro would normally just smolder, but in the pure oxygen environment at higher pressure, it essentially exploded.

The final design flaw that prevented the astronauts from surviving was the hatch on the spacecraft. The Block I hatch was a two-piece affair, an inner hatch, which opened inward and fit into the fuselage like a cork in a bottle. When pressure built up inside the spacecraft, the inner hatch had become wedged in. In addition, it was secured in place by bolts that took time to remove. There was also an outer hatch that was unwieldy and time consuming to remove. These and other elements had conspired to place Grissom, White, and Chaffee inside a tiny, explosive crematorium. There was no way for them to get out in time to avoid death.

There was plenty of blame to go around. The assembly of the Command Module by North American was proved to be shoddy, as evidenced by the faulty wiring and other problems. The hatch design, dictated by NASA, was a poor choice for an emergency departure from the spacecraft. The use of Velcro was inadvisable, at least in that abundance. Finally, the choice of a pure oxygen atmosphere at that pressure was pure insanity, and while North American had suggested that testing with a pure oxygen environment was a bad idea in at least one memo, NASA had insisted it was safe enough. As fellow Apollo astronaut Frank Borman later said at a Congressional inquiry, they had suffered from a "failure of imagination" that resulted in the death of a crew.[19]

The Command Module was already in redesign for the Block II version, and to this process was added an entirely new, one-piece hatch that opened outward and could be released in less than ten seconds. The wiring inside the capsule was redesigned and relocated. The use of Velcro and other flammable materials were minimized. The space-suits, which had also caught fire, were augmented with flame-retardant outer cloth. Finally, and perhaps most critically, the spacecraft would never again be pumped up to sea-level pressures with pure oxygen—a complex system was designed to use a normal, nonflammable Earth atmosphere when at lower altitudes—whether during a test or in flight—that was later purged and replaced with the lower pressure 5 psi pure oxygen once they were near orbit.

Once again, lessons had been learned, but this time at the cost of three lives, and Kranz and his charges at Mission Control would never forget that.

**Figure 4.5. Kranz speaks with Apollo astronauts Jim McDivitt (*left*) and Rusty Schweickart (*center*), during the Apollo 5 unmanned test flight in January 1968. (Courtesy of NASA.)**

The first manned flight of Apollo, Apollo 7, was launched on October 11, 1968. Wally Schirra (a Mercury veteran), Walt Cunningham, and Donn Eisele (both rookies) rode into orbit in the thoroughly redesigned Block II Apollo capsule atop a Saturn-IB. After just shy of eleven days in space, the crew returned to Earth—the mission had been another success for NASA's march to the moon. By this time, and partly because of the increased administrative needs involved in recovering from the fire, Kraft had been moved out of direct mission operations and was now the Director of Flight Operations. Kranz was left in charge of Mission Control, and remained a flight director as well.

Then came a change or plans for Apollo 8.

There were more incremental test flights planned for the Apollo system, but with Kennedy's end-of-the- decade deadline looming, a number of these missions seemed superfluous to key members of NASA's senior management. In addition, the Lunar Module was facing huge challenges at Grumman, its manufacturer—it was overweight and had technical problems. The LM had become a pacing item, and it was holding back the test schedule. By midsummer 1968, change was in the wind, and it would be dramatic.

"Come August, I get a call to go over to Kraft's office, and Kraft says, 'Sit down. I want to talk to you,'" Kranz recalled. "'I'm taking a team over to Huntsville today, and what we're going to do is we're going to propose going to the moon this December.' Well, this is in August, so we're talking in four months we're going to go to the Moon. And this sort of catches everybody by surprise here."[20] If it caught Kranz and his team by surprise, it was an absolute stunner outside their ranks. NASA was proposing to skip a whole series of test flights and send astronauts into orbit around the moon, using the Command/Service Module that had been tested just once in Earth orbit. And because there would be no Lunar Module on this flight, they would be severely bending one of their own rules—with just the single engine on the Service Module to get them home, Apollo 8 would have an enormous single point of failure. If that engine failed to fire once the crew was in lunar orbit, there would be no coming home. As Pete Conrad, who would command the mission of Apollo 12 in 1969, said about the engine, "If it doesn't work we're just going to become the first permanent monument of the space program," circling the moon forever with a dead crew.[21] Dark humor aside, it was a sobering thought to everyone involved.

Kraft asked Kranz to let him know by the following morning if there were any compelling reasons not to fly this risky mission. "Obviously you can pull together a list of reasons you can't do things as long as your arm, but that wasn't the way you did things in those days. What you really did is say, 'Why can't we?' So you kept looking for the opportunities that were there," Kranz said.

After a brief but intense few days of looking over all the parameters and variables, assessing the risks, and tying up just about every mainframe computer at the Johnson Space Center to compute trajectories and launch dates, they had their answer. Apollo 8 was go.

It was at about the time that Kraft put Kranz in overall charge of the Flight Control Division—this duty would be in addition to his role as flight director. He was, therefore, enormously busy tackling his expanded duties when Apollo 8 launched in December, 1968. This was a very different experience for him and must have given him a taste of what it took for Kraft to step away from being a flight director.

"This is one where I was almost glad I was sitting on the sidelines," Kranz remembers:

> It always seems that the people who are watching the mission get more emotionally involved in the mission than the people who are doing it, because the people who are doing it have got to be steely-eyed missile men, literally. I don't care whether they're 26 years old or 35. The fact is that you've got to stay intensely focused on the job. It is the people who are sitting in the viewing room [a glassed-in area behind the control center], I think, who have it the toughest, [the off-duty] flight directors who are trying to find a way to plug into somebody's console so they can listen in to what's going on.
>
> But I think that was probably the most magical Christmas Eve I've ever experienced in my life, to actually have participated in a mission, provided the controllers, worked in the initial design and the concept of this really gutsy move, and now to really see that we were the first to the moon with men. We were at the point where we were setting records, literally, in every mission that we flew in those days, because the Russians had long since ceased to compete, it was obvious that we had the best opportunity for the lunar goal. And this was just a magical Christmas. I mean, you can listen to Borman, Lovell, and Anders reading from the Book of Genesis today, but it's nothing like it was that Christmas. It was literally magic. It made you prickly. You could feel the hairs on your arms rising, and the emotion was just unbelievable.[22]

Apollo 8 made ten orbits of the moon before heading home on December 25, 1968. The gamble had paid off, and with the LM nearing readiness at Grumman, it was just a matter of months before they would attempt the first lunar landing. But there was little time for celebration, there was no opportunity to bask in the victory. There was much work to be done before NASA could proclaim Kennedy's goal accomplished.

Apollo 9 was an Earth orbital mission only, designed to test the LM's systems in flight. It launched on March 3, 1969, and for the next ten days, the crew ran a thorough shakedown of all the systems needed for a lunar landing. They performed multiple rendezvous and docking exercises between the LM and the Command Module. Then they fired the LM's descent engine, and after they separated from the lower stage, fired the ascent engine, a separate power plant in the upper stage of the LM:

> This was also a good opportunity to really continue this testing of this lunar module, in five days. It was a nine-day mission, and the five days that we were working with the lunar module, we did 10 engine burns, 10 maneuvers, so we did essentially two major maneuvers a day. At the same time, we'd do rendezvous with the spacecraft, actually put the crew in this lunar module, and if we couldn't get the two back together, this crew wasn't coming home, because there's no heat shield on it.[23]

As it turned out, the practice they got with the LM would pay off just over a year later, when Apollo 13 got into trouble. After one of the countless simulated mission sessions they held after Apollo 9, the Simulation Supervisor called Kranz and his team onto the mat:

> One day my team didn't do the job right, and when we were debriefing the training, our SimSup, which is our training boss, comes to us in the debriefing and says, "Why did you leave the lunar module powered up? Why are we using all that electrical power? Don't you think you should have developed some checklist to power this thing down? Whenever you've got trouble, you ought to find some way to conserve every bit of energy, every bit of resources you've got, because some day you might need it." In debriefing—the guy's name was Jerry Griffith—and we had to say, "Jerry, that's a good idea. We weren't doing our job. We weren't thinking. We were thinking too many other things." So we started developing a series of emergency

power-down checklists that really was our first line of defense when Apollo 13 came along.[24]

Apollo 10 was cued up next—the final practice run for the Apollo spacecraft system before the lunar landing of Apollo 11. The crew would fly to the moon, and then two of the astronauts would enter the LM, leaving one behind in the CM, and descend to within relative spitting distance of the moon—without landing—before staging the LM and flying back to orbit to dock with the CM and come home.

"The final thing that we had to do was conduct a full-blown dress rehearsal, and that's exactly what we did on Apollo 10. It's to now put all of these pieces together and do a rehearsal for the lunar landing, including making a low pass across the surface of the moon. This came off incredibly well," Kranz recalls.[25]

It must have been frustrating to the mission commander, Tom Stafford, and his LM pilot, Gene Cernan, to get so close to the moon and not grasp the prize of landing there (Cernan would do so on Apollo 17, the final lunar mission), but they had a job to do, and they did it well. The only major hiccup was that when they staged the LM at about 47,000 feet in altitude, the LM's upper stage suddenly spun wildly out of control—the radar had been loaded with the wrong program for ascent. It only took a few moments for Stafford and Cernan to regain control and continue their ascent to the waiting CM, but it had been a close thing.

Finally, in July 1969, came the big one. The grand prize. The culmination of a decade of intensive work, double shifts, vast amounts of overtime, and strained family lives. The time had come to attempt the first landing on the moon.

While the crew at Mission Control was a cohesive, cooperative team, Kranz could not help but feel competitive about this first landing attempt. There had been a quiet competition among the flight directors to excel during the simulations—each team wanted to be on point for the first landing. They each did their best, and all were qualified for the assignment in unique ways. Soon the announcement came from Cliff Charlesworth, the lead Flight Director for the Apollo 11 mission, and the man who would dole out the assignments. Kranz recalls it as being almost anticlimactic. Charlesworth wandered into Kranz' office, stalled a bit for effect—they both knew why he was there—and said, "You're going to do the lunar landing." Kranz recalls, "I just ricocheted around the office virtually all day, and I don't think my secretaries ever saw me as happy."[26]

Though it would be hard to intensify the experience they were all having, the training for the landing attempt, just months away, was a whole new level of stress.

The training process for Apollo 11 began very late, because the lunar landing software in the simulators wasn't ready. Charlesworth had supervised launches before, and the Saturn V had proved to be reliable, so he got through his training as quickly as possible to allow the other Flight Directors access to the simulators—there were only so many hours to go around with the simulators, the computers, and Mission Control, where the ground-based half of the simulations were conducted. "That gave [Glynn] Lunney and me pretty much access to the simulators when we needed to in the months of May and June, because we were going to be launching in July, and we generally cut our training off about two weeks prior to launch because the crew had other things to do, we had other things to do."[27]

"So, the first month of training with my team went like a champ," Kranz recalls. "We were familiar with the lunar module, we had a hot hand, too cocky."

The SimSup running the sessions sensed their cockiness, and decided that it was time to deflate the team controlling the lunar landing a bit. "He started increasing the pressure associated with the descent phase," Kranz said:

> Now, when you're going down to the Moon, just like landing an airplane, there is essentially a dead man's box. No matter what you do, you can throttle up, you can change your attitude, you're going to touch the ground before you start moving back off again, and it's the same kind of condition, but it isn't a neatly drawn line; it's a set of variables, and it depends upon the altitude and the speed at which you're descending down how this box is defined.[28]

In short, if you were too close to the surface and a problem occurred, the LM could crash. Add to this the delay in sending and receiving radio messages to the moon, and the fact that this was still largely unknown territory—they had only experienced a lunar landing in computer simulations—and it was clear that mistakes could still be made. The SimSup decided to toss Kranz's team into the grinder:

> SimSup really laid it to us, and it related to this dead man's box and this lunar time delay. We went through a series of scenarios that were almost—it seemed like forever. It was only a couple of weeks, but it seemed a lifetime where we could not do anything right. Everything we would do, we would either wait too long and crash or we would jump the gun and abort when we didn't have to, and the debriefings were absolutely brutal during that period of time.[29]

It got so bad that Kranz was losing confidence in his own abilities, and the Apollo 11 crew was to frustrated to even speak to the Mission Control team for a time.

The launch date was alarmingly close, but, after more grueling simulations, Kranz finally felt it was coming together. Then, another crowbar was thrown into the simulations.

"The final training runs, invariably, are supposed to be confidence-builders," Kranz said. "It's to the point now this is the last time you're going to have an opportunity—generally things are going to go right during the course of the mission, so let's stay within the box, let's build the confidence of this team, etc."[30]

But the SimSup, Dick Koos (who had been with NASA since the beginning), didn't see it that way. As Kranz recalls, "We started our final day of training, and about midway through the day, we had done more aborts, and . . . it was starting to get irritating to me, because what I wanted to do was practice the landing, continue to refine the timing of the landing."

Koos had other plans:

> I think it was either the last or second to the last training exercise . . . and midway through the descent training, we saw a series of com-

puter program alarms. We'd never seen these before in training. We'd never studied these before in training. My guidance officer, [Stephen] Steve Bales, looked at the alarms and decided we had to abort.

We aborted, and I was really ready to kill Koos at this time, I was so damned mad. We went into the debriefing, and all I wanted to do is get hold of him at the beer party afterwards and tell him, "This isn't the way we're supposed to train," and in the debriefing we thought we'd done everything right.

Koos comes in to us, and he says, "No, you didn't do everything right. You should not have aborted for those computer program alarms. What you should have done is taken a look at all of the functions. Was the guidance still working? Was the navigation still working? Were you still firing your jets? and ignored those alarms. And only if you see something else wrong with that alarm should you start thinking about aborting." We told him he was full of baloney.[31]

But, ever diligent, Kranz told Bales to work up a list of computer error codes, perhaps as much to get Koos off his back as to prepare for any possible issues in the upcoming landing—though Kranz was nothing if not thorough. Bales dutifully complied, and the list sat at Bales's console, unloved and unwanted, until the day of the actual landing.

It looked like this:

Rule 5-90, Item 11, powered descent will be terminated for the following primary guidance system program alarms—105, 214, 402 (continuing), 430, 607, 1103, 1107, 1204, 1206, 1302, 1501, and 1502.[32]

Unloved though it might be, it came in handy.

Apollo 11 launched on July 16, 1969, with Neil Armstrong in command, Michael Collins as the Command Module pilot, and Buzz Aldrin as his Lunar Module pilot. The last designation is a bit of a misnomer, as in the Apollo program the mission commander actually piloted the LM with the LM pilot backing him up. Within hours of reaching orbit, Collins had pulled the Command Module free of the third stage of the Saturn V, turned 180 degrees, linked up with the LM, and pulled it out of its protective housing. The three men sped off to the moon, LM in tow, for their rendezvous with destiny.

Apollo 11 arrived in lunar orbit on July 19. Though it was only the third time a crewed spacecraft had flown behind the moon, then fired its engine to stay there, tensions were far lower than they had been for Apollo 8. The hardware had proven itself reliable, and besides, the media's attention was on the upcoming landing attempt; even those in Mission Control were affected.

Kranz and his "White Team" controllers had a thirty-two-hour break before the landing—the other Flight Directors would handle the activities until then, and everyone wanted to make sure the guys on the consoles were at their best for the upcoming events. On landing day, Kranz went through his usual morning activities to prepare for work—discipline was his mantra and routine his shelter. When he got into his car, personal tradition took over.

"I pump myself up each time I get ready to do something. I can hear 'Stars and Stripes Forever,' by John Philip Sousa," he remembers. "At this time also we had eight-track players, so I had them in the car. Every place I'd go, I'd have John Philip Sousa. And this is the way I get up to speed, get the energy, get the adrenaline flowing."[33] Kranz had Sousa at home, Sousa in the car, and Sousa in his office at Mission Control. Years later he would be invited to guest conduct Sousa before a major orchestra. He was just that kind of a guy—patriotic to a fault and corny to the core. It's a wonder the astronauts didn't have to listen to Sousa during the descent to the moon.

Upon arriving at Mission Control, Kranz hung up his jacket and put on his vest—his wife had made him an extra special one for this day. "Marta had made me a silver and white brocade vest, very fine silver thread running through this thing," he recalled.[34] He had been feeling a sense of predestination while heading into the room. He felt like a military leader:

Patton has always been my favorite, because Patton felt that . . . he had been in the battlefields of Thermopylae, he had been with the

Roman Legions, he'd been fighting at Sparta, he had this feeling of predestination. Well, I've always had the same feeling. It's sort of weird. But basically you walk down this hall in Mission Control, and again, I'm not thinking of a lunar landing, I just feel that myself and the team I've got, from the time that we were born, we were meant for this day. And it's funny how these things feel.[35]

Once he was "vested" and inside Mission Control, the impressions shifted. "You can tell people have been in there for a long period of time. There's enough stale pizza hanging around and stale sandwiches and the wastebaskets are full. You can smell the coffee that's been burned into the hot plate in there. But you also get this feeling that this is a place where something's going to happen. I mean, this is a place sort of like the docks where Columbus left, you know, when he sailed off to America or on the beaches when he came on landing."[36] Kranz felt destiny just around the corner.

He looked at his controllers, just filing in to take over from the outgoing shift, Glynn Lunney's group. His team—he called them "White Flight"—was now in place. The faces were young—many looked as if they had not yet had their first shave, though he knew that most were in their mid-twenties. At thirty-five, Kranz was the old man in the room. When asked why he selected such a young crew to run the mission, he responded, "I wanted men who had never known failure."[37]

Kranz checked in with his controllers, then went to his console to get settled in. Soon, 240,000 miles away, Mike Collins was undocking the CM *Columbia* from the LM *Eagle*, and then Armstrong rotated the LM so that Collins could check to assure that the landing gear was locked in place. Within minutes he had given Armstrong a verbal thumbs up, and Armstrong and Aldrin headed to a lower orbit to begin their descent.

The landing attempt would begin shortly—a decade of preparation was on the line—and it was time for Kranz to address the troops. Though his speech was not recorded, this is what he remembers saying:

Okay, all flight controllers, listen up. Today is our day, and the hopes and the dreams of the entire world are with us. This is our time and our place, and we will remember this day and what we will do here always.

In the next hour we will do something that has never been done before. We will land an American on the moon. The risks are high . . . that is the nature of our work.

We worked long hours and had some tough times but we have mastered our work. Now we are going to make this work pay off.

You are a hell of a good team. One that I feel privileged to lead.

Whatever happens, I will stand behind every call that you will make.

Good luck and God bless us today![38]

He also recalls adding: "Today we will either land, abort, or crash attempting to land. The last two outcomes are not good."[39]

Then Kranz ordered the doors to the control room locked, and the circuit breakers locked in place—he would not risk a repeat of the blackout from the Gemini days. He called this preparation condition "battle short"—a term from his air force days—indicating protection against any potential interruptions he could guard his team from.

Kranz now had a decision to make—should *Eagle* begin its descent to the lunar surface? Were they ready? Everything seemed okay on Armstrong's end, but Kranz had a problem—the radio signal from the LM to Houston was terrible. Not only were the voice comms garbled, but the telemetry—a second radio channel that transmitted the condition and status of the systems aboard the LM via a digital signal—was all but useless.

"We can't communicate to them; they can't communicate to us," he recalls. "The telemetry is very broken. We have to call Mike Collins in the Command Module to relay data down into the lunar module, and immediately this mission rule has come into mind because it's decision time, go/no go time."[40]

Kranz stalled a bit to see if the comms would improve, and in the LM, Aldrin fiddled with the directional antenna to try and get a better signal.

"It just continues, broken, through about the first five minutes after we've acquired the data, but we get enough data so the controllers can make their calls, their decisions," Kranz says.[41] "We move closer now to what we call the 'powered descent go/no go.' This is where it's now time to say are we going down to the lunar surface or not. Now, I have one wave-off opportunity, and just one, and if I wave off on this powered descent, then I have one shot in the next rev[olution] and then the lunar mission's all over. So you don't squander your go/no go's when you've only got one more shot at it." The room was tense—the radio signal was still awful. "Come right up to the go/no go, and we lose all data again. So I delay the go/no go with the team for roughly about forty seconds, had to get a data back briefly, and I make the decision to press on; we're going to go on this one here. So I have my controllers make their go/no go's on the last valid data set that they had." He was asking them to look at the frozen display on their screens—they would make a decision based on what the LM's status had been before the signal dropped had out again.

He polled his team of controllers, and each said "Go." In fact, young Steve Bales was so keyed up, he fairly screamed it . . . "GO!" You can hear Kranz chuckle over the intercom; it broke the tension for a bit.[42]

Kranz gave CAPCOM Charlie Duke, an astronaut who would fly on Apollo 16, instructions to tell the Armstrong and Aldrin that they were go for powered descent. Duke also radioed the instruction to Collins to relay to his comrades below.

*Eagle* started its powered descent—the engine fired to reduce speed and allow them to slowly fall toward the lunar surface, 50,000 feet below them.

Within minutes more problems found them, "Like flies to a picnic lunch," as Kranz put it.[43] The guidance officer realized that the *Eagle* was coming down "long"; they would overshoot their prescribed landing zone. "Instead of being at the landing point we had planned, we're now moving further down range to the edge of our landing footprint, which is very rocky," Kranz said.[44]

He allowed the landing to continue.

Then new problems arose: "We've got this new landing area that we're going into, we're fighting the communications ... and now a new problem creeps into this thing, which is this series of program alarms."[45]

The computer in the LM had locked up, and instead of displaying the altitude and speed, and indicating that it was continuing to function, it had stopped showing anything except "1202." This was an error code ... but which one? Was it critical? Should they abort?

CAPCOM Duke muttered, "It's the same one we had in training ..."

Bales grabbed his once-unloved list of computer error codes assembled after the recent simulation in which he had called an abort and gotten reamed for it. He scrambled to see if "1202" was on the list, as did his supporting engineers in the "back room," an area down the hall where other engineers and technicians provided support for the controllers.

Thankfully, "1202" was not on the abort list. Bales said, "We're Go on that alarm. If it doesn't recur, we are Go."[46]

With the computer acting up, they could not see the information from the radar to tell them how high they were, but just as Bales gave the call to continue, the radar data returned.

"So we go through this kind of an exercise at the same time we're accepting this radar data. We tell them we're [go on] the alarms, we tell them to accept radar, go on the alarms, you know, radar's good, getting close ... we're continuing to work our way down to the surface."[47]

They had eight minutes to go.

The computer alarms recurred, shifting from a "1202" to a "1201," but it was in the same class of problem. Bales told Duke to let the crew know that Mission Control would monitor the radar data on the ground and keep him apprised.

But the problems were not over. As *Eagle* passed 7,000 feet, Armstrong realized that the area below them was far rockier and more heavily cratered than expected—they were coming down on the far side of the smooth area they had planned to land on. He soon arrested his descent and began flying horizontally, searching for a place to set down that would not jeopardize their ability to take off when it was time to return to orbit. Any obstacle—a rock, crater rim, or depression—more than about eighteen inches in elevation could doom them.

Fuel was now becoming critical.

Kranz said, "CAPCOM, we are go for landing,"[48] and Duke relayed this to Armstrong and Aldrin. But Kranz was acutely aware that fuel was low—and within moments this was confirmed by a call from a controller, who said, "Low level." In the simulations they had always landed by now, but Armstrong was still scooting above the surface, scanning the terrain ahead. Aldrin was calmly reading off the altitude and velocity numbers, gently encouraging Armstrong to start heading down again.

"Armstrong has to pick out a landing site, and he's very close to the surface. Instead of moving slowly horizontal, he's moving very rapidly, and 10 and 15 feet per second," Kranz said. "We've never seen anybody flying it this way in training. Now [controller] Carlton calls out '60 seconds,' and we're still not close to the surface yet, and now I'm thinking, okay, we've got this last altitude hack [measurement] from the crew, which is about 150 feet, which now means that we've got to average roughly about three feet per second rate of descent, and I see he's at zero. So I say, 'Boy, he's going to really have to let the bottom out of this pretty soon.'"[49]

Duke was still relaying information to the crew, until Deke Slayton, the Mercury astronaut who was now in charge of the Astronaut Office and was standing next to Duke, punched him in the shoulder—hard, as Duke tells it—and said, "Shut up and let them land."[50] Duke sheepishly complied. The room fell quiet . . .

They had a half a minute before the crew was supposed to abort the landing if they could not find a place to set down.

"Now we're 30 seconds off the surface of the Moon, and . . . incredibly rapidly I go through the decision process. No matter what happens, I'm not going to call an abort. The crew is close enough to the surface I'm going to let them give it their best shot," Kranz says. "At the same time, the crew identifies they're kicking up some dust, so we know we're close, but we don't know how close because we don't know at what altitude they'd start kicking up the dust, and then we're to the point where we're mentally starting, waiting for the 15-second call, and Carlton was just ready to say, '15 seconds,' and then we hear the crew saying, 'Contact.'"[51]

One of three five-foot metal bars extending from the landing legs had touched the surface, and the LM had settled onto the moon. It was over.

Almost.

The room erupted in cheers, and the people in the viewing area behind Kranz were going crazy with glee and relief. Kranz looked down at his hand . . . he had broken the pencil he was clutching sometime in those last dramatic minutes, and his hand ached. He marveled for a moment, then said, "Alright, settle down in here!" The control room quieted quickly, but nobody could erase the smiles.

Then the next problem popped up. A gauge indicating pressure in a fuel line was climbing rapidly and quite unexpectedly. The cold from the lunar surface had caused fuel in a pipe to freeze, and the pressure behind it was building. This had never come up in a simulation, and the controller responsible for the spacecraft systems, as well as the Grumman and TRW team in the back room (TRW was an aerospace contractor that had built the descent engine) were watching it closely. If it got too high, it would either blow a relief disk (an assembly built for just such a contingency), or rupture the fuel line or cause a tank to explode—the latter could endanger the crew.

Kranz recalled, "Again, one of the things you can never test, the heat soak-back from the engine and the surface now is raising the pressure in that bottle very dramatically, and now we're wondering if this damned thing's going to explode and what the hell are we going to do about it."[52] But Kranz had studied every last system in the LM down to the minute details, and he knew that there was a relief system installed in the fuel lines. "If the pressure got so high, it actually blows the disc and the valve, rather than blowing the bottle up."

About the time he was pondering a stay/no stay decision, and whether he might need to issue an order to abort the mission and order them to return to orbit, the ice plug melted and the pressure dropped. His work for today was done.

We get handed over to Charlesworth's team, and I'm going over to the press conference with Doug [Ward, the public relations official] and it was the first time, actually, you really had the chance to unwind and think about, "Today we really landed on the Moon." It's the goddamnedest thing you'd ever seen in your entire life ... you were right there, you were doing all of these things, but every American went through their thing, and we were only limited to a second where we could really imagine and be happy with what we did. It was an incredible feeling.[53]

Kranz went to his press conference, told the world a brief accounting of the tale, then hung around Mission Control for the EVA. He went home a very happy man that day, and was able to enjoy a brief respite as a spectator until it was time for Apollo 11's return to Earth. The rest of the mission went off without a hitch, a testament to fine engineering, relentless training, and absolute dedication.

Figure 4.6. Kranz leans on his console during the Apollo 11 EVA. (Courtesy of NASA.)

**Figure 4.7. Gene Kranz with the "White Team" from Apollo 11. Kranz is to center left. (Courtesy of NASA.)**

Apollo 12 followed later that year, launching on November 14, 1969. It was another nearly flawless mission. There were glitches to be sure—the Saturn V and Apollo spacecraft were struck by lightning twice during their ascent, but the rocket continued into orbit safely. Then, and far less critically, the TV camera burned out when it was accidentally pointed into the sun during their moonwalk, but that was a small thing—America's second lunar visit was, again, a triumph.

Then came Apollo 13. This story is well known due to the extreme drama of the emergency, and the movie *Apollo 13*. But Kranz's perspective bears retelling.

To recap: the Apollo 13 mission launched on April 11, 1970. In the commander's seat was James Lovell, a veteran of the Gemini program as well as Apollo 8. Fred Haise, a rookie, was his Lunar Module pilot. Jack Swigert was the Command Module pilot, stepping in at the last moment for Ken Mattingly, who had tested positive for rubella (German measles) a week before the launch date. The crews and NASA management had to scramble to make the swap, but Swigert had been the designated backup and had trained along with the rest of them, so he was ready to go. But nobody liked last minute crew changes.

Everything was going well until fifty-six hours into the mission. At that time, when the astronauts were 205,000 miles from Earth, there was an explosion aboard the spacecraft—a large oxygen tank on the Service Module, the power and propulsion unit behind the Command Module, had ruptured. This was the primary source of both life support and power for the spacecraft, which was now crippled.

Kranz was the lead Flight Director for this mission, and his extensive and hard-won skills were taxed to the limits and beyond to get his crew home:

[Apollo] 13 was, again, a mission where the basic maturity of this team . . . just spread forth in almost a magnificent fashion. We had made the decision missions earlier that we would always have four Mission Control teams in place during the course of a mission. And this gave us several advantages, because quite frequently the mission events don't fit neatly into eight-hour shifts. So, a team might have to do what we called a "whifferdill." Either show up a shift early or show up to a shift late. And having the fourth team in position made that transition much easier.

But it also was designated as a crisis team; that if we had any problems during the course of a mission, major problems, this team would try to find some way to work itself off line and the remaining three teams . . . would continue to work eight-hour shifts throughout the mission, whatever it turned out to be. My team was designated as lead team; and . . . our principle responsibilities during the mission: we were going to be doing the lunar orbit insertion and also we were going to do the ascent from the Moon. And that's what we had been trained to do. During the course of the mission, it changed dramatically.[54]

The crew had just completed a TV transmission back to Earth, and were going through the last of the evening's procedures prior to a rest period. They worked through the list as it was read up from mission control through the final checks.

"We were down to the final entry," Kranz recalls, "and we advised the crew that we wanted a cryo stir."[55] The fuels in the service module were cryogenic—that is, ultra-cold liquids stored in highly insulated tanks—but they were so cold that they would get slushy, so electrical fans were installed inside the tanks and powered up periodically to keep the fuels in a liquid state. The fan inside one of the oxygen tanks on Apollo 13 was damaged, however—the tank had been dropped many months before, and although it had been tested and seemed to be okay, a very small bit of insulation had been stripped off one of the wires powering the fan. When Swigert switched it on, there was a tiny spark. It was almost like a replay in miniature of Apollo 1—an electrical spark inside a pure oxygen environment. The oxygen ignited, and the tank exploded.

Kranz remembers it vividly: "All of a sudden I get a series of calls from my controllers. My first one is from guidance. It says, 'Flight, we've had a computer restart.' The second controller says, 'Antenna switch.' The third controller says, 'Main bus undervolt.' And then from the spacecraft I hear, 'Hey, Houston, we've had a problem' … Within Mission Control, literally nothing made sense in those first few seconds because the controllers' data had gone [to] static briefly; and then when it was restored, many of the parameters just didn't indicate anything that we had ever seen before."[56]

The calls coming from the controllers and those from the crew were increasingly concerning, but trying to figure out what they actually meant was vexing.

"Finally, the training that's given the controllers kicked in," Kranz recalls.

Sy Liebergot was on the EECOM (environmental and electrical control). Power, tank pressure, water reserves—most everything critical to crew survival was under his purview. And he was suddenly not having a good day. "None of the data Sy is seeing, from his standpoint, is believable. Very quickly it looks like we've lost one of our fuel cells and possibly a second one." The second oxygen tank was zero—it was the one that blew—and the pressure in its twin, the first tank, was falling.

Kranz estimates that it was about five minutes into the emergency before his thoughts began to crystallize—he needed to get selective in his listening and pull things together. He barked at his controllers: "'Okay, all you guys, quit your guessing. Let's start working this problem.' Then I use some words that sort of surprised me after the fact. I say, 'We've got a good main bus A. Don't do anything to screw it up. And the lunar module's attached, and we can use that as a lifeboat if we need to. Now get me some backup people in here and get me more computing and communications resources.'"[57]

After twenty more minutes of watching the only remaining large oxygen tank drain, Kranz got a call from EECOM. "Liebergot then comes to me and says, 'Hey, flight, I want to shut down fuel cells 1 and 3.'" Kranz remembers. "And I say, 'Sy, let's think about this.' And he says, 'No, flight, I think that's the only thing [that's] going to stop the leaks.' And then I go back to him the third time and I say, 'Sy . . .' and he says, 'Yeah, flight, it's time for a final option.'"[58] Kranz trusted his people implicitly, but it was asking a lot. That would kill the power generating fuel cells, and they would be on the LM's batteries for the remainder of the flight, which were not designed to operate for anything like that length of time.

"By this time, Lovell's called down and indicating they're venting something. And we've come to the conclusion that we had some type of an explosion onboard the spacecraft; and our job now is to start an orderly evacuation from the command module into the lunar module. At the same time, I'm faced with a series of decisions that are all irreversible. At the time the explosion occurred, we're about 200,000 miles from Earth, about 50,000 miles from the surface of the Moon," Kranz says. They would have to make a decision about how to get the crew home; the landing was obviously off the table now. "During this period, for a very short time, you have two abort options: one which will take you around the front side of the Moon, and one which will take you all the way around the Moon."[59]

Flight Director Lunney had polled the relevant team members and took the news to Kranz. Kranz recalls, "If I would execute what we call a 'direct abort' [around the front side of the moon] in the next 2 hours, we could be home in about 32 hours. But we would have to do two things: we'd have to jettison the lunar module, which I'm thinking of using as a lifeboat, and we'd have to use the main engine [the large rocket on the Service Module]. And we still have no clue what happened onboard the spacecraft. The other option: we've got to go around the Moon; and it's going to take about 5 days but I've only got 2 days of electrical power. So, we're now at the point of making the decision: which path are we going to take? My gut feeling, and that's all I've got, says, 'Don't use the main engine and don't jettison this lunar module.'" Some of his engineers were concerned that the explosion might have damaged the main engine, and it could also explode, which would probably kill the crew. "And that's all I've got is a gut feeling. And it's based, I don't know—in the flight control business, the flight director business, you develop some street smarts. And I think every controller has felt this at one time or another. And I talked briefly to Lunney, and he's got the same feeling."[60]

Kranz made the call—they would continue past the moon and figure out how to stretch the life support and power resources. Aboard the stricken spacecraft, the crew concurred.

Kranz now needed to break down what activities would be assigned to which teams to figure out what to do to get his crew home safely. Arnie Aldrich, who had been with NASA since the beginning, was given overall responsibility for maintaining the master checklists. Electrical engineer John Aaron would oversee consumables—power, breathable oxygen, and water. Bill Peters, an LM controller, was assigned the task of configuring the LM as a lifeboat.

Aaron said, "There's no way we're going to make 5 days with the power in the lunar module. We got to cut it down to at least 4 days, maybe 3." So Kranz split the team further. "I had one team working

power profiles. I had another group of people that was working navigation techniques. I had a third one that was integrating all the pieces we
need. My team picked up the responsibility to figure out a way to cut
a day off the return trip time ... once a person was given the responsibility to do the job, everybody would snap to and support him. Once
decisions were made, you never second-guessed those decisions."[61]

This process continued for the first twenty-four hours, at which
time they fired up the LM's descent engine to get Apollo 13 headed
home at a faster clip, and then powered down as many of the spacecraft's systems as was possible. The crew was only allowed to use 200
watts—about a quarter of what a modern microwave oven requires.

Just as they accomplished this, the carbon dioxide levels started
to rise, as so well documented in the *Apollo 13* movie: "The crew was
suffocating. We had to invent techniques of using the square chemical scrubbers we used [in] the command module and be able to
adapt those over to the lunar module." It was much harder than you
might think to instruct a tired, shivering crew to build the jury-rigged
scrubber adaptors in time, but they did it. One more problem licked.

As Apollo 13 closed in on Earth, there was one more major hurdle to
be overcome: configuring for reentry. "We had a command module that
was our reentry vessel. It had the heat shield, but it had only about two
and a half hours of electrical power lifetime. We had the service module,
which is where the explosion had occurred; it was virtually useless. We
had the lunar module, [which] was attached on the other end of this
stack through a small tunnel, and that was our lifeboat," Kranz says. "We
had to come up with a game plan to move this entire stack into an attitude where we could separate all three pieces in different trajectories
so they wouldn't collide with each other in entry."[62] If they simply separated one craft from the other without some planning, they could drift
into each other and spoil the chances for a successful reentry.

Kranz went on:

> Then the crew had to evacuate from the lunar module lifeboat at the
> very last moment, power up the command module, get its computer

initialized, separate the pieces, and get into attitude for entry. So, this is the game plan we were coming up with. And we didn't really get all the pieces put together and get them verified in simulators until about ten hours prior to the time that we had to execute this plan.[63]

Not much of a margin.

We got the procedures up to the crew. Jack Swigert had the command module part of the procedures. Fred Haise had the lunar module. And about the time we were voicing up these procedures, we realized how desperate it was onboard the spacecraft. It was in the high 30s, low 40s. The crew [was wearing] the cotton coverall flight suits they had.[64]

To make matters worse, Haise had some kind of infection and was running a temperature of about 104 degrees F, deep in space and far from any medical intervention; it was a flight surgeon's nightmare made real.

Undocking from the LM went off perfectly, and the Command Module plunged into Earth's atmosphere. With a number of Apollo flights behind them, Mission Control knew how long it should take to complete the reentry, and all they could do now was wait. Apollo 13 was engulfed in a fireball, barreling through the atmosphere, and radio communications were impossible:

Each controller during blackout—this is an intensely lonely period. Because . . . the crew's on their own. And they're left with the data that you gave them, maneuver data, attitude information, all of these kind of things. And each controller's going back through everything they did during the mission and, "Was I right?" And that's the only question in their mind.[65]

The controllers' eyes would dart from their control screens, hoping to see the return of data transmission, to the mission time clock to see how close they were to getting their answer: had the crew survived?

And when it hits zero, I tell [CAPCOM Joe] Kerwin ... "Okay, Joe, give them a call." And we didn't hear from the crew after the first call. And we called again. And we called again. And we're now a minute since we should've heard from the crew. And for the first time in this mission, there is the first little bit of doubt that's coming into this room that something happened and the crew didn't make it. But in our business, hope's eternal, and trust in the spacecraft and each other is eternal. So, we keep going.[66]

It was a full ninety seconds beyond the time they would have expected to regain communications when the answer came in. Apollo 13 was home; the crew was safe.

"Our celebration always started with cigars. I don't know what the young controllers are going to do ... today, because you can't smoke in Mission Control," Kranz mused. "But anyway, you start with the cigars. And they've got to be good cigars, because nobody in Mission Control is going to ... smoke a bummer. And we had some darn fine cigars!"[67]

**Figure 4.8. Gene Kranz enjoys a "darn fine" cigar after the successful conclusion of the Apollo 13 mission. (Courtesy of NASA.)**

After the cigars, the controllers wrapped up the necessary details and left Mission Control, their job done. It was time for some serious drinking, and a well-deserved rest. "This was an honest-to-God broth-erhood that existed in those days that I don't think anything, any group of people, in peacetime has ever come together in a similar fashion," Kranz says.[68]

There were four more lunar landings in the Apollo program, and then a year of comparative downtime during which they trained and prepared for the Skylab program. Skylab repurposed the upper stage of a Saturn V as a space station, and that mission was a stunning success in its own right. There were problems, the largest of which was damage suffered by the space station during launch (see chapter 7). But this was overcome with the same intensive preparation and professionalism that had served them so well during the long march to the moon, and Skylab was a spectacular success, hosting three crews, the last of which stayed on the station for eighty-four days.

Apollo had one more act before it was consigned to history, however—the 1975 flight of Apollo-Soyuz, in which an American Apollo spacecraft docked and orbited with a Soviet Soyuz space-craft. This was a far cry from the adventure of lunar exploration, but it required the same close attention to every detail that any previous flight had.

Then the teams of Mission Control went off to other duties, or left the program altogether, as NASA prepared for the inauguration of the space shuttle, which would not fly into orbit until 1981. Six years was forever in this business, and Kranz saw the denouement coming long before the end of the Apollo-Soyuz mission. Near the end of the Apollo 17 flight, the fact that the end was near had begun to creep into his awareness.

"I was interested in the legacy," Kranz said; that is, what it meant to work in Mission Control, which had been Kranz's life for the past twelve years. "I wanted to leave a different legacy than the one Kraft [had left]. Kraft had established the legacy of the flight director. I was looking at the one—the legacy in a broader sense—the one of the team. The one of the Mission Control itself."[69]

Robert McCall was an artist who had been in NASA employ for

most of the space race. Kranz always admired McCall's ability to look at a scene in Mission Control, or an image on one of the monitors, and within a minute make a pencil sketch of what he saw that communicated not just the imagery but the *feeling* of what was taking place.

"So, I asked Bob to design us an insignia for Mission Control, and ... I said I wanted to talk about the commitment." Kranz and McCall talked about what it meant to serve in Mission Control, to take the lives of astronauts into your hands, help them to achieve a nation's goals, and then bring them home safely. The final result was stunning, and is used to this day as a mission patch and emblem for Mission Control.

**Figure 4.9. Kranz confers with Chris Kraft (*right*) during the inaugural launch of the space shuttle in 1981. (Courtesy of NASA.)**

"It represents everything we learned about spaceflight, the commitment and the teamwork of the Mercury and the Gemini Programs," Kranz recalls. The emblem personified "the mission. Morale, believing so strongly in your mission, your team, and your success that you literally cause the right things to happen. Tough and competent came out of the Apollo fire, where basically we weren't tough enough. We didn't step up to our responsibilities. We have to remember, in the business we're in, we're always accountable for what we do or what we fail to do. Competent [is something that] we can never stop learning. So basically, I sketched out to Bob the elements that I wanted to be representative of the emblem of Mission Control. And he agreed to go do this."[70]

The results of that commitment to teamwork, to excellence, to perfection, continue through this day as Mission Control prepares to team up with private industry to send astronauts to the International Space Station and beyond, to once again return to the moon. And Kranz could not be prouder of the results.

**Figure 4.10. Kranz during the shuttle mission STS-26 in 1988. (Courtesy of NASA.)**

Gene Kranz continued to work with NASA through much of the shuttle program, retiring in 1994. He lives in retirement in Webster,

Texas, with his wife, Marta. His six children visit him often, as do his countless grandchildren. But it would be a mistake to think of him as a Barcalounger-bound duffer, for his "retirement" has consisted of writing an autobiography and working a busy lecture schedule—he is a highly valued speaker on leadership across the United States and worldwide.

**Figure 4.11. Kranz speaking at the Marshall Space Flight Center in 2006. (Courtesy of NASA.)**

And he is still a patriot. At the end of a 2005 interview that I conducted with him, when we had gone through my long list of questions, which he answered patiently and with unfailing good cheer, I asked him if he had anything to add. At this point, most people say, "No, I guess that's about it." But not Kranz. He paused for a moment, staring at the floor between us. Then he looked up, and I could see the determination borne of a lifetime of service to a cause that mattered, to a country he loved, and said, "What America will dare, America can do."

I heard the strains of Sousa's "Stars and Stripes Forever" in my head.

# CHAPTER 5

# MARGARET HAMILTON: THE FIRST SOFTWARE ENGINEER

**Figure 5.1. Margaret Hamilton in 1995. (Courtesy of Wikimedia Creative Commons, author: Daphne Weld Nichols, licensed under CC BY-SA 3.0.)**

On July 20, 1969, the mission of Apollo 11 was well underway, with Neil Armstrong and Buzz Aldrin headed toward the moon's surface in the Lunar Module. Apollo 8 had orbited the moon in December 1968, and the flight of Apollo 10 had tested the LM in a "dry run" pass earlier that year, in May, flying near the lunar surface without landing. With these goals met, two astronauts were now descending to a landing site in the Sea of Tranquility, a process that would require many thousands of things to go right in order to be successful, and could fail if only a handful of things went wrong.

One of those potential showstoppers would be a failure of the new navigation computer installed aboard the LM, which would guide it most of the way to its landing site. This was new technology and software, and it pushed the limits of computer science well beyond where it had been just a couple of years earlier, especially in miniaturized flight-rated computers. In short, it *had* to work.

It was midday in Houston, and at the Johnson Space Center controllers were hunched over their consoles in Mission Control, watching systems and subsystems on the LM as it began its historic descent toward the moon.

Many minutes earlier, Flight Director Gene Kranz had given the nod to Charlie Duke, Apollo astronaut and the CAPCOM for the landing—he would be the voice link to Armstrong and Aldrin during their landing attempt. Radio contact had been sketchy, but Kranz gave the go-ahead and committed to the first phase of the landing: "*Eagle*, Houston. If you read, you're go for powered descent. Over," Duke had said.[1]

The crew acknowledged the communique. Guiding them down was a small computer. At a time when mainframes filled rooms and computer data was saved on paper punch cards and reels of magnetic tape, the Apollo Guidance Computer (AGC) was a miracle of miniaturization. Aerospace contractor Raytheon had built it, and had managed to shrink what should have taken up a large closet into a unit little larger than a fat briefcase. It had an unremarkable thirty-six kilobytes of memory— about one-tenth that of a 1990s floppy drive—but that was enough to get the Apollo spacecraft from Earth orbit to lunar orbit, down to the surface, back to lunar orbit and then home. Key to this was well-designed, highly compact software. That came from the computer lab over at the Massachusetts Institute of Technology, or MIT. A small team there had been laboring over the project for a few years now, and tonight would be the acid test of their work. They were gathered in a contractor's room at Mission Control, where engineers and technicians could watch the mission and be available for input if needed.

Margaret Hamilton was one of the members of that team. In fact, she was in charge of a big part of the software project. She was in her early thirties now, but she had been in her mid-twenties when she was hired to work in a computer lab at MIT—a lab otherwise staffed almost exclusively by men; this was the "Mad Men" era after all—and she had been central to much of the work. Hamilton and the others in attendance were listening intently to the radio downlink. Up there, 240,000 miles away, flew a tiny, fragile spacecraft with two men aboard, making humanity's first attempt at a lunar landing. Lives depended on her work and the efforts of countless others ... but in the next few minutes, it would be her work that was central to success or failure.

Gene Kranz had summed up the day less than an hour before when talking to his flight controllers. As he later put it, "From this point forward, we would have one of three outcomes. We would land on the Moon, we would abort while trying to land, or we would crash. The last two were not good."[2]

Hamilton had not heard the words, but she knew that the computer aboard the LM could be the difference between those outcomes.

By now, Armstrong was well aware that he was off target. They were coming in long and would overshoot their projected landing zone by miles. The computer did not know this, however, and was steering them to unwelcome territory. He was adjusting as best he could. "Our position checks down range show us to be a little long," he said calmly, as they descended. In reality, he was looking at a lot of ugly craters and car-sized boulders that could kill the fragile LM. In Houston, people tracking the navigation in the back rooms had their charts out, trying to figure out about where the *Eagle* might be and where the astronauts could set down. Fuel was being rapidly depleted.

As Armstrong and Aldrin were descending past about 38,000 feet, the radio had been pretty quiet—Aldrin was reading off occasional bits of data from the computer, and the calls were coming up from Mission Control to keep the astronauts apprised of their status. Then, in firmer, terse tones, Armstrong said, "Computer alarm."

That dumped cold water on everyone—the computer couldn't fail now! The appropriate controllers were squinting at their consoles, trying to make sense of the message. After a few moments, Armstrong spoke again, reading off the computer alarm code. "It's a 1202," he said, and when that was met by silence from the ground, he added, "Give us a reading on the 1202 Program Alarm."[3]

At that moment, the landing was in doubt. Nobody on the consoles immediately knew what that computer alarm meant. Technicians were flipping through stacks of notes, trying to see what it might mean. Time was of the essence—if the computer quit on them, Armstrong might have to abort the landing, an inherently dangerous maneuver in itself.

At that moment, Margaret Hamilton's programming work hung in the balance. She had written that bit of computer code, along with the other hundreds of thousands of lines of code she had contributed, and the next few minutes could see her greatest victory or one of the worst moments of her young life.

Hamilton had been born in Paoli, a small town in southern Indiana, in 1936. Paoli was a heavily Quaker community (her grandfather had been a Quaker minister), and she attended a Quaker liberal arts college called Earlham, graduating with a degree in mathematics in 1958. While still in school, she worked as an actuary at the Traveler's Insurance Company, and then, after getting married upon graduation, she taught high school math and French to support her husband, who was attending Harvard. The couple moved to Boston, where Hamilton intended to attend Brandeis University for a graduate degree in abstract mathematics, a highly theoretical branch of the discipline.

But, as so often happens, life (aided in this case by the birth of a daughter) got in the way, and as her husband continued his studies, aiming for a law degree, Hamilton took a job working for a professor of meteorology at the Massachusetts Institute of Technology (MIT), creating programming for a computer that would be used to predict weather patterns. People in the field called what she was writing "software"—a term that had only been in use for two years at the time.[4]

Hamilton's tendency to think and live independently of social mores of the time did not take long to rise to the surface: "When my husband was in law school, they wanted the 'law wives,' my being one of them, to pour tea. And I said to my husband, no way am I pouring tea! As a Harvard law wife, if I go to Harvard Law School, fine, I'll do what the men do, but I'm not going to be put in that position, and he was very proud of me, that I had taken that stand."[5]

**Figure 5.2. One of the massive SAGE computer banks ca. 1962. (Courtesy of NASA / United States Air Force.)**

Having proven to be proficient in the somewhat arcane art of writing computer code, in 1961 Hamilton moved across campus to MIT's Lincoln Laboratory, their nascent computing center. While NASA had begun using many women as "computers"—human calculators, executing long and complex calculations by hand—there were fewer women in the software field, and at MIT Hamilton worked almost exclusively with men. The project she was now assigned to was called the Semi-Automatic Ground Environment (SAGE) program for the US Air Force. The program was tasked with identifying enemy aircraft flying toward the United States. The computers it ran on filled large rooms—in fact, each computer weighed 250 tons, and there were twenty-four of the massive machines.[6]

Hamilton said that project was the kind of work that newcomers were often assigned to at the lab:

> What they used to do when you came into this organization as a beginner, was to assign you this program which nobody was able to ever figure out or get to run. When I was the beginner they gave it to me as well. And what had happened was it was tricky programming, and the person who wrote it took delight in the fact that all of his comments were in Greek and Latin. So I was assigned this program and I actually got it to work. It even printed out its answers in Latin and Greek. I was the first one to get it to work.[7]

That was the kind of mind she had—always thrilled with a new challenge.

In 1963 Hamilton moved to another lab at MIT, the Charles Stark Draper Laboratory. This lab specialized in defense applications and, more recently, spaceflight. She would now, at twenty-eight years old, be working on computer software for a new NASA project called Apollo, for which the computers themselves were still under development.

Hamilton rose through the ranks quickly due to her ability to grasp the complex algorithms involved and put them into practice, and she was soon in charge of the effort to design navigation programs for the computers that would fly in the Apollo Command Module. A big part of the job was making the software usable for the extensive simulations that NASA would carry out to train both the astronauts and the technicians in Mission Control. "Systems simulations were a mix of hardware and digital simulations of every—and all aspects of—an Apollo mission which included man-in-the-loop simulations, making sure that a complete mission from start to finish would behave exactly as expected," she later said.[8]

It was in this role—it was now the mid-1960s—that she came to understand what was at stake: "The space mission software had to be man-rated. Not only did it have to work, it had to work the first time. Not only did the software itself have to be ultra-reliable, it needed to be able to perform error detection and recovery in real time. Our languages dared us to make the most subtle of errors. We were on our own to come up with rules for building software. What we learned from the errors was full of surprises."[9] The term "man-rated" meant that humans would be flying on these missions, and any errors in the computer code could cost lives. It was a weighty responsibility for a young woman from Paoli, Indiana.

One of Hamilton's interests early on was error detection. All software has bugs—you've seen them in action if you ever ran Windows 2000 on your computer. She started asking around to see what people were doing about this—surely someone smart had dealt with the problem. Her colleagues told her they used the "Auge Kugel" method. Having no idea what this meant, she continued to investigate. "I wanted to know what this Auge Kugel method was. I found out much later on that it meant 'eyeballing' in German," she said in 2001.[10] Essentially, the programmers would sift through millions of lines of code to see if anything popped out: "That's what they kept using, and that's how they discovered certain errors." When a software error was written up for record-keeping, the designers would simply write "bug."

That method didn't sit well with Hamilton, and it got her interested in finding ways to discover errors using the computer itself, a quest that would pay vast dividends just a few years later.

She was soon put in charge of the software design team for both the Command Module and the Lunar Module, and she began investigating the commonality between the two sets of programming for the different spacecraft. One, the Command Module, was designed to fly from the Earth's surface, through space, into lunar orbit, and then back to Earth. The other, the Lunar Module, was intended to take the astronauts to a predetermined landing site on the moon and then back up to the orbiting Command Module. Both sets of software had to have the ability to rendezvous with an orbiting spacecraft as well. There turned out to be plenty of overlap, and Hamilton took advantage of this while continuing to design new programs.

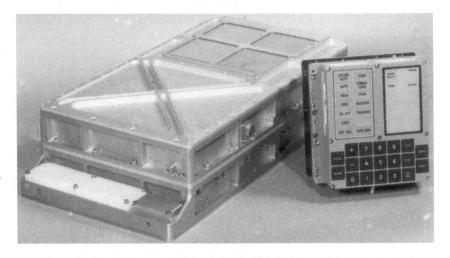

**Figure 5.3. The Apollo Guidance Computer. To left is the processor enclosure, to right the DSKY (Display-Keyboard) interface unit. It was the smallest and most powerful computer to fly at the time. (Courtesy of NASA.)**

Around this time, Hamilton coined the term "software engineer" to describe her work (she was listed as "Programming Leader" inside the code itself).[11] It will not surprise you that this too was a field dominated by men, and while Hamilton got along with her male co-workers, she had to make sure that her work was always above reproach.

By now, Hamilton was the parent of a young daughter, Lauren.

These were the days when mothers were expected to stay home to raise their young children, and she was often called out by other women for not properly raising her daughter. "People used to say to me, 'How can you leave your daughter? How can you do this?'" she recalls.[12] Having no other recourse, she took Lauren to work with her, which proved to be not only a good bonding exercise but may also have saved the lives of the Apollo 11 astronauts as well.

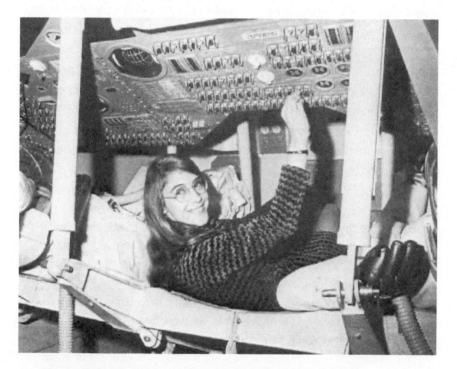

**Figure 5.4. Hamilton testing software inside an Apollo Command Module simulator. (Courtesy of NASA/MIT.)**

Lauren had seen her mother climbing in and out of an Apollo capsule simulator and, at age four, thought it would be fun to "play astronaut." Hamilton allowed her to climb into the simulator during quiet times while she worked on her code. During one such playtime, Lauren entered a gibberish code into the DSKY (for "display and key-board")—the computer display terminal in the Command Module, which was still running. The computer immediately crashed, losing all its navigational data. Hamilton became curious as to how this had happened, and she set aside her other tasks to identify the specific problem. She discovered that her daughter had inadvertently loaded a computer program called P01, a software routine designed to be used in pre-launch activities, at the same time that the computer had been running another program that it would use during flight toward the moon. The conflicting commands had caused it to lock up completely.

Hamilton discussed this with her bosses at MIT and was told not to worry about it. She then raised the issue with NASA, who said that adding even more software code to deal with this potential problem, which they did not really see as likely to occur, could cause more problems than it solved. They added that the astronauts were "highly trained," and that this software conflict would never happen in flight. To appease her concerns, they added a note in the computer operation instructions that said "Do not select P01 during flight."[13]

Hamilton later recalled,

> One of the things I remember trying very hard to do was to get permission to be able to put more error detection and recovery into the software. [This way] if the astronaut made a mistake, the software would come back and say "You can't do that." But we were forbidden to put that software in because it was more software to debug, to work with. So one of the things that we were really worried about is what if the astronaut made a mistake—we were also told that the astronauts would never make any mistakes, because they were trained never to make mistakes.[14]

Despite the admonishment, she developed error detection and recovery programming—in essence, programs that would monitor other programs as well as the functioning of the computer. These routines had to deal not only with the machines but also with possible errors committed by the people running the computers 240,000 miles from Earth. Part of her solution was to allow the computer to interrupt input from the human operators by flashing a priority display, a numeric code, that would alert the user that the computer was dealing with a programming issue, and to give it time to work through the tasks it had to in order to prevent it from crashing.

As it turned out, astronauts *did* make mistakes. In fact, two years later one would make the exact same mistake, with the same codes, in the same situation, as Lauren had.

During the flight of Apollo 8, the first mission sent to the moon to orbit for a day before returning to Earth, Jim Lovell was loading programs into the computer while en route to the moon. He committed the same error as a four-year-old girl had earlier, entering program P01 into the computer midflight. This not only brought the computer to a standstill, but it wiped clean the navigation data that Lovell had been collecting and entering into the computer. Hamilton received an urgent call at MIT—NASA needed a fix, and they needed it *now*; they had astronauts in jeopardy. Nine hours later, after sifting through an eight-inch stack of programming code printed on paper, Hamilton presented NASA with a solution. The new data was uploaded to the Command Module's computer and the emergency was over. But this was just a harbinger of what would occur during the Apollo 11 landing.

It bears mention that the message "Do not select P01 during flight"—the instruction that was supposed to keep the programming problem from occurring—had been right there all along, in the program itself and in the instructions for users. Unsurprisingly, it hadn't made one bit of difference.

By early 1969, and with the first Apollo lunar landings coming up in a few months, the pressure was on. Hamilton now oversaw 350 programmers and engineers who were working on the project.

It's worth noting that changing the programs in the spacecraft computers involved more than just altering a few numbers on printouts and punch cards (a paper card that had holes punched in it to convey binary code to a mechanized reader attached to a computer). Much of the programming was embedded on large "memory cards." These slabs of plastic were covered with thousands of tiny wires that ran through small ferrite beads, or "cores," and the system was called "core rope memory." The cores were little bits of magnetized metal with tiny holes bored through their centers. The choice of running a wire through the hole, or alternatively past it, indicated whether the binary signal coming out the other end was a "1" or a "0." It was a primitive, though incredibly robust, way to store programming—there was no way this part of the code could be lost or disrupted.

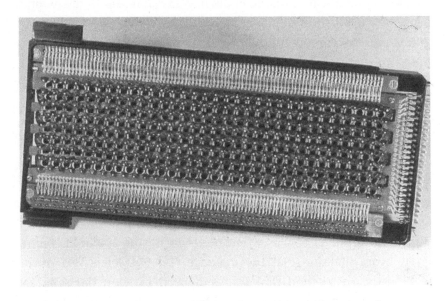

**Figure 5.5. An example of core rope memory as used on the Apollo flights. (Courtesy of NASA.)**

A group of workers at the subcontractor were fabricating these memory cards for the Apollo computers, and most of them were women who had previously worked in textile mills. They were found to be extremely capable at this intricate work, threading wires not much larger than a human hair through the tiny ferrite beads. All of this was done under microscopes. The resulting product is almost artistic in appearance, and hundreds of these cards were fabricated in near clean-room conditions.

But because this programming was literally "hard wired," whenever there was a change in the program the affected card would have to be sent back to the manufacturer and the offending wires rerouted to reflect different binary code—individual wires would either be pulled out of the little metal beads, or would be fed through them. It was an incredibly time-consuming and expensive process, so designing clean code as early as possible was highly desirable. It's not hard to see why the effort to create the Apollo Guidance Computer software is estimated to have taken over 1,400 total work-years to accomplish.[15]

Hamilton and her team soldiered on, preparing for the flights that would lead up to the first landing attempt. In March 1969, the Apollo 9 mission flew in Earth orbit, and was the first crewed mission to carry both a Command Module and Lunar Module. This mission, which would be the first manned test of the LM, required it to rendezvous and dock with the CM multiple times. This was the first real trial by fire of Hamilton's guidance software working in harmony in both spacecraft.

Then, in May, Apollo 10 flew to the moon. The LM separated from the CM in lunar orbit, and then descended to within about eight and a half miles of the lunar surface but did not land—this was the "dry run" for the mission of Apollo 11. Everything worked well until the crew staged the Lunar Module to return to orbit—the ascent stage separated from the descent stage and fired its own rocket engine to ascend. Right at that moment, the ascent stage began to roll violently, a result of the crew having punched duplicate commands into the guidance computer. The crew disengaged the computer and manually regained control, completing the mission safely. While there were worries about the computer programs for a short time, the engineers soon realized that the problem was a matter of user error—the software was okay.

The decks were clear for Apollo 11 to depart Earth in July.

As Neil Armstrong and Buzz Aldrin neared the surface of the moon, Hamilton's computer error code popped up—four stark green digits replacing the altitude and velocity code on the computer's display with "1202." When Armstrong asked Mission Control to "Give us a reading on the 1202 alarm," controllers scrambled to decipher the error code. A computer alarm could be cause for Armstrong to abort the landing, which would involve an emergency abort maneuver that was itself dangerous. Kranz needed an answer—right now.

Steve Bales, the twenty-four-year-old controller on the guidance console, was desperately scanning a crib sheet of computer alarms he had prepared as a result of the countless simulations they had endured. It was up to him to make the call to abort or continue. Within moments he heard a voice through his headphones—it was another engineer, Jack Garman, speaking from the "back room," where a technical support team worked to back up the flight controllers. Garman had a similar list and remembered Hamilton's 1202 code from a simulation just days earlier, where they had faced a simulated computer error. Garmin assured Bales that it was okay to continue.

Hamilton heard all this from another room. She had not had time to intervene, but she could imagine what was going on. For some reason the guidance computer had overloaded and flashed the 1202 alarm—it had too much to do. As per her software design, this mean that the computer was disregarding nonessential tasks and returning to its core problems of computing and relaying speed and altitude to the systems that were needed continue flying the Lunar Module toward the designated target on the Sea of Tranquility far below.

The alarms continued to sound, but Bales assured Kranz that all was well, with CAPCOM Duke relaying the call, "Same type, we're go." The landing continued successfully.

Upon later investigation it became apparent that once again human error had caused the potential emergency. Aldrin had, per the checklist, left a switch in a positon that fed data continuously from the rendezvous radar, which would be used to fly back to the Command Module in the event of an abort, even as the landing radar, which was looking at the moon below—fed data into the computer as well. The resulting cascade of data was too much for the small computer and caused it to become overloaded, triggering the alarm as it prioritized its tasks. Had Hamilton's interrupt not been built into the software, as NASA had originally suggested, the computer would likely have crashed and, had an abort not been successful, the Lunar Module could have followed suit.

While few realized it immediately, Hamilton was the hero of the hour. Her insistence on putting the emergency interrupt programming into the guidance computer had saved the landing. And it all started with a four-year-old girl "playing astronaut."

"After the manned missions, I guess I personally just had a sense of history about wanting to not just remember things, but to do something based on what lessons were learned," Hamilton recalls. "Just like when an error happened, you'd find a way to not let it happen again."[16] She decided to stay at NASA to work on Skylab, America's first orbiting space station, and then on the space shuttle program. Both of these incorporated her software design in their computer software, which was based on the programming used for the Apollo missions.

After her work for NASA, Hamilton opened her own company, Higher Order Software, and further developed designs for error prevention and fault tolerance based on her NASA experience. In 1986 she formed her own consulting firm, Hamilton Technologies, where she developed Universal Systems Language, which was designed to detect and resolve errors in computer code as early in the development process as possible. This work, and the resulting products spawned by it, revolutionized software design.

Hamilton has been widely recognized for her contributions to spaceflight and computer programming, receiving numerous awards and honors in the decades since the Apollo program. In 2016, Hamilton received the Presidential Medal of Freedom, the highest of civilian honors, in recognition for her important work.

Hamilton is also a staunch advocate of enabling women in the workplace, and she has observed that in many ways it's actually harder for women to break into the software field today than it was in the 1960s. At that time, and as a woman, she had been an anomaly in a brand-new industry, and it was the "Wild West," as she has called it— there was room for rapid change (and for women) in the nascent field of software design.[17] Now, in a more established and mature phase, the field can be more exclusionary. And it's not just software, it's a broader problem, she says.

"I gave a talk at my college, at Earlham, where they wanted to know what it was like being in an engineering field and being a woman. And it almost seems to be worse in some ways today than it was back in the early days. When you see women not being allowed to drive in certain countries, or you see that women can't become priests ... When you start seeing this ... every single one of those things impacts our culture, or impacts women or minorities as to whether they can even do something or not," she said in a 2017 interview.[18] "Until we start making changes, until our leaders stop admiring people who do things that encourage that, we have a problem."

Hamilton continues to inspire women worldwide with her frequent talks and interviews. Her story of forging new pathways in the earliest days of software design still resonate today, over fifty years later.

**Figure 5.6. Hamilton in 1989. (Courtesy of NASA.)**

# CHAPTER 6

# NEIL ARMSTRONG AND BUZZ ALDRIN: "FIRST MEN"

On July 20, 1969, a frail-looking machine slowly made its way down to the lunar surface from a sixty-mile orbit. The Lunar Module (LM) was an odd contraption, appearing like a thirty-foot-tall abomination of lumps and bumps—the ascent stage looked positively diseased—with the lower, or descent, stage covered in rumpled Mylar foil. But spindly though it looked, this machine had a single purpose, one that it would fulfill brilliantly: to land two men on the moon and, later, return them to orbit.

At 20 hours, 17 minutes, 40 seconds UTC (Universal Coordinated Time, also called Greenwich Mean Time), or 2:17 p.m. Houston time (where Mission Control was located), the LM *Eagle* touched the lunar surface, settling onto the edge of the landing zone in the Sea of Tranquility. It had been a harrowing descent, with computer alarms threatening to scuttle the landing, and then a desperate scramble to find a safe place to set down while fuel supplies dwindled to almost nothing.

Then, at just under eighteen minutes past the hour, two men landed on the moon for the first time. They had trained together, launched together, flown through 240,000 miles of darkness together, and worked together—as a team—to cross that final sixty miles from lunar orbit to the "magnificent desolation" of Tranquility. And inside the LM, standing in their pressure suits, they arrived at precisely the same moment with a gentle thump.

**Figure 6.1. The Lunar Module as it appeared shortly before descending to the moon's surface on July 20, 1969. (Courtesy of NASA.)**

There was really no "first man" on the moon, and Neil Armstrong would be the first to agree. There were two of them, Armstrong and Buzz Aldrin, who made the final crossing to the surface. They had been delivered to the proper place to begin the final phase of that landing by a third teammate, Mike Collins, who orbited silently above. So, a team of three flew to the moon, and two landed there, on Apollo 11—backstopped by over two dozen other astronauts and hundreds of thousands of men and women who labored tirelessly for a decade, against overwhelming odds, to get them there—in a sense, they were all "first" to the moon. But it was Armstrong, of course, who took that first step onto *Luna incognita*.

Neil Alden Armstrong was born on August 5, 1930, in the small town of Wapakoneta, Ohio. But that was just a jumping-off place for his family, and they lived in sixteen different towns while he was growing up—his father was a state-employed auditor whose assignments kept the Armstrongs mobile. But the family endured, and by 1944 they were back in Wapakoneta, where Armstrong completed high school.

He became enamored of flying very early in his life, when airplanes were just starting to be made of metal and had a single wing. In areas like rural Ohio, there were still cloth-covered biplanes in service, and many were relics of World War One.

When he was six, Armstrong had his first airplane ride, in a Ford Trimotor. The plane was a noisy beast, with loud, radial engines (in which the pistons are arranged in a circle instead of inline like auto engines)—one on each wing and another on the nose of the aircraft. The fuselage was covered in corrugated metal, which looked as if it could have been peeled from the roof of an old metal shed. The passenger seats were usually made of wicker—a woven fibrous plant prized for its strength and, in this case, its light weight. The planes of the era did not inspire the highest confidence by modern standards, but Armstrong was just thrilled with the sense of altitude and speed. These machines *flew*, and Armstrong's fascination with flight grew from that time forward.

As he advanced through his early teen years, Armstrong read the books of Charles Lindbergh, who had flown solo across the Atlantic in 1927, and those of Lindbergh's wife, Anne Morrow Lindbergh, and this further deepened his passions for the air.

As he grew older and progressed through school, however, his love for the idea of flight was shouldered aside somewhat by a fascination with how flight was accomplished—the study of aeronautical engineering. He realized that the idea of piloting was almost secondary to his core drive:

I began to focus on aviation probably at age eight or nine, and inspired by what I'd read and seen about aviation and building model aircraft, why, I determined at an early age—and I don't know exactly what age, while I was still in elementary school—that that was the field I wanted to go into, although my intention was to be—or hope was to be an aircraft designer. I later went into piloting because I thought a good designer ought to know the operational aspects of an airplane.[1]

In his teens, Armstrong also took flying lessons at the Wapakoneta airfield, receiving his pilot's license when he was sixteen years old. But his adventures were not limited to the air ... Armstrong's interest in mathematics and physics led him down a few unconventional pathways during those years—in the form of school projects. He recalled,

One [project] was building a Tesla coil. I think it was probably about a 50,000 volt Tesla coil, good enough to light up fluorescent bulbs in the next room. Then a wind tunnel. ... My knowledge of aerodynamics was not good enough to match the quality of the Wright Brothers' tunnel, [though] at that point I suppose I was equally educated to them. But it was a fun project. [It] blew out a lot of fuses in my home. Because I tried to build a rheostat which would allow the electric motor to change speed and then get various air flows through the tunnel, not altogether successfully.[2]

Armstrong was also an avid member of the Boy Scouts of America, moving up through the ranks to Eagle Scout—the top tier—and remained engaged with the scouts in one way or another over much of his lifetime.

By the time he graduated from high school, World War Two was over, but that did not stop him from signing up for military service. Armstrong received a college scholarship to Purdue University, where he studied aeronautical engineering. His coursework included flight training and other subjects, in addition to engineering. It was intended to be a seven-year commitment—quite a stretch to an eighteen year old. "Two years of [university study], then go to the navy, go through flight training, get a commission, and then serve in the regular navy for a total then of three years of active duty, after which the plan would be to return to university and finish the last two years," Armstrong later said.[3] But in 1949, the navy called him to active service early due to the Korean War.

Armstrong, Neil A., ENS., USNR, 505129/1315
23 May 1952   80-G-679736

**Figure 6.2. Neil Armstrong as a young navy officer in 1952. (Courtesy of the United States Navy.)**

Armstrong reported to Naval Air Station Pensacola in Florida for training in a propeller-driven WWII aircraft. A year later he was in Texas and had qualified in carrier landings, the most challenging of flight assignments. He then transitioned to the F9F Panther, a fighter jet, and in mid-1951 he was headed to war-torn Korea. Within a week of his first flights there his jet encountered an antiaircraft trap—a

system of cables slung high in the air to ensnare American aircraft—
and part of his right wing was sheared off:

> I actually ran through a cable, an antiaircraft cable, and knocked off
> about six or eight feet of my right wing. If you're going fast, a cable
> will make a very good knife.... I was flying on the wing of John Car-
> penter. He was an air force major, on an exchange program with us.
> We talked it over and decided [that I should] not to try to land it,
> because if I got a little bit too slow and started to snap, I would have
> no [ability] to control it after that, so consequently decided it would
> be better to jump out. So, took it down south into friendly territory
> and jumped out in the vicinity of Pohang Airport, K-3, which was
> operated by US Marines.[4]

**Figure 6.3. The F9F Panther fighter jet Armstrong flew over
Korea. His aircraft is in the background. (Courtesy of the United
States Navy.)**

After finishing his tour in Korea, Armstrong returned home to
complete his coursework at Purdue. Then he got his first job out of
school at NASA's predecessor, the National Advisory Committee for
Aeronautics (NACA), at the Lewis Research Center in Cleveland, where
he worked as a research pilot:

The flying involved doing work with new anti-icing systems for air-craft, which we had a C-47 (or R-4D or DC-3) with various kinds of anti-icing equipment that we would fly out in the worst weather we could find out [over] Lake Erie and try to pick up a lot of ice and find out which were the most efficient ways of shedding it.[5]

He also flew airplanes that would fire small research rockets:

This project involved flying an F-82 [fighter jet] ... to high altitude and launching a multi-stage rocket downward into the atmosphere to get very high Mach numbers at very low altitudes, and therefore very high heat transfer rates.[6]

Then he would write-up the results of the research:

The only product of the NACA was research reports and papers. So when you prepared something for publication, either as a principal or associate author of some sort, you had to face the "Inquisition," which was the review of said paper by experts who were predom-inantly lady English teachers or librarians who were absolutely unbearably critical of the tiniest punctuation or grammatical error, and that is what NASA needs today.[7]

After his time at Lewis, he was transferred to Edwards Air Force Base in California, the Mecca for test pilots. Once there, he flew a variety of high-performance and experimental aircraft, including some of the early X-planes—the experimental jets that filled the gap between the X-1 (the rocket plane that first broke the sound barrier) and the X-15, NASA's ultimate rocket plane. The X-15 was amazing, but very challenging to fly, and completely experimental, so problems were usually part of the process. One flight was intended to test a new system that was supposed to automatically help stabilize the X-15 as it transitioned from the airless edge of space back into an atmosphere—within minutes the X-15 could be transitioning from Mach 1 to Mach 5, for example. By automatically integrating the reaction control system (the maneuvering jets for use outside the Earth's atmosphere) with the aerodynamic controls (such as elevons and the rudder, for use within the atmosphere), this system was supposed to help the pilot make a relatively seamless transition back into normal aerodynamic flight. Part of this system was supposed to limit the g-forces a pilot would have to endure during the reentries. In the flight testing this system—dramatized in the opening scenes of the movie *First Man*— Armstrong's career as an aviator was almost ended:

> We [had] tried this many times in the simulator without any dif-
> ficulty, but when we really did it in flight, I couldn't [quite achieve
> 5 g's], so I [kept] pulling to try to get the G limiter to work. In the
> process, I got the nose up above the horizon. We'd done this in the
> simulator, never had any problem with it. But I found when I did it

**Figure 6.4. Neil Armstrong poses with the X-15. (Courtesy of NASA.)**

in real flight, I was actually skipping outside the atmosphere again. I had no aerodynamic controls. That was not a particular problem, because I still have reaction controls to use, but what I couldn't do is get back down in the atmosphere ... I [rolled] over ... and tried to [drop back into] the atmosphere, but [the aircraft] wasn't going down because there was no air to bite into. So I just had to wait until I [fell low enough] to have aerodynamic control and some lift on the wings, [then] immediately started making a turn back. But by that time I'd gone well south of Edwards.[8]

Due to his reentering downtrack from the predetermined spot, he would, by some accounts, make a descending turn far above the Rose Bowl in Pasadena, some eighty miles to the south of Edwards, to head back. The X-15 was unpowered at this point—just a stubby-nosed glider—and did not have much lift. Reaching the flat desert landing zone while he still had some altitude was critical:

It wasn't clear at the time I made the turn whether I would be able to get back to Edwards. That wasn't a great concern to me because there were other dry lakes available there. I wouldn't want to go into another one, but I certainly would if I needed to. [Eventually], I could see that we were going to make it back to Edwards, so I landed without incident on the south part of the lake.[9]

That's some flight test.

As his work on the X-15 continued, Armstrong was being segued into another air force program, project Dyna-Soar. Dyna-Soar, also later called the X-20, was to be an early space shuttle–type spacecraft—one that would launch atop a Titan III booster (a more powerful variant of the rocket than launched the Gemini spacecraft) and orbit the Earth with one astronaut/pilot at the controls, with a possible variant that would carry a total of five in cramped quarters. Upon deorbiting, Dyna-Soar would also glide back to a landing strip at Edwards Air Force Base to land on wheels and skids, much like the X-15 from which it was derived. It was planned to be reusable up to ten times. This was a purely military project, designed to deploy spy satellites and perform orbital reconnaissance on the Soviet Union and other adversaries of the United States.

**Figure 6.5. Period artist impression of Dyna-Soar being launched atop an Atlas rocket, which would have propelled it only into suborbital test flights. (Courtesy of the United States Air Force.)**

The Apollo Guidance Computer's interface unit, or DSKY, in 1969. It was a miracle of miniaturization for its time. *Image from Flickr: Steve Jurvetson, licensed under CC BY 2.0.*

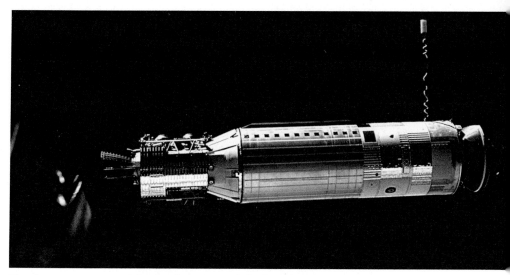

The Agena stage used for rendezvous and docking practice during the Gemini flights. This is the view that greeted Neil Armstrong as he flew the Gemini capsule toward a docking—and then, all hell broke loose. *Image courtesy of NASA.*

John Glenn is helped inside the tiny Mercury capsule in preparation for his orbital flight on February 20, 1962. *Image courtesy of NASA.*

An official NASA portrait of John Glenn in 1961. *Image courtesy of NASA.*

Buzz Aldrin near the American flag during the Apollo 11 moonwalk. *Image courtesy of NASA.*

Buzz Aldrin checks out the Lunar Module during the flight of Apollo 11. *Image courtesy of NASA.*

Pete Conrad before the flight of Apollo 12, doubtless making a joke for the photographers. *Image courtesy of NASA.*

The Fallen Astronaut statue and associated plaque left on the lunar surface by the Apollo 15 astronauts. *Image courtesy of NASA.*

Yuri Gagarin, the first human to fly in space.

Gene Kranz, sporting his Apollo 17 splashdown vest in 1972. *Image courtesy of NASA.*

Margaret Hamilton poses with the paper printouts for the Apollo navigation and guidance software in the mid-1960s. *Image courtesy of NASA.*

The liftoff of John Glenn's Mercury flight on February 20, 1962. *Image courtesy of NASA.*

The liftoff of Apollo 11 on July 16, 1969—one of the seminal photographs of the space race. *Image courtesy of NASA.*

The crew of Apollo 11. *From left:* Mike Collins, Buzz Aldrin, and Neil Armstrong. *Image courtesy of NASA.*

Pete Conrad (*left*) tweaks Gordon Cooper after the flight of Gemini 5. *Image courtesy of NASA.*

The Vostok 6 spacecraft after landing in the Soviet Union in 1963.

An official portrait of Valentina Tereshkova before her Vostok flight. *Image from Alexander Mokletsov / RIA Novosti Archive, licensed under CC BY SA 3.0.*

Skylab as seen in 1973, after Pete Conrad and his crew had salvaged the space station by effecting repairs to the solar panel. *Image courtesy of NASA.*

Dyna-Soar was in development from 1957 through 1963, at an ultimate cost of about $660 million (or about $5.3 billion in current dollars), but it never flew—the program never got beyond the mockup stage. But before Dyna-Soar was canceled, Armstrong had a decision to make. On May 25, 1961, just a few weeks after Alan Shepard had made his landmark fifteen-minute suborbital flight, President Kennedy had made an announcement to the Congress that riveted the nation. He declared that the United States "should commit itself to achieving the goal, before this decade is out, of landing a man on the moon and returning him safely to the Earth."[10] Armstrong, like so many others, was stunned by the speech.

Subsequent to this, NASA issued a call to expand the corps of astronauts from the original seven selected for Project Mercury, adding another nine for the new Apollo lunar program. Armstrong applied, but he did so quite close to the deadline and his application arrived a full week after the closing date. As fate would have it, however, an air force associate of his, who was not working for NASA, spotted Armstrong's name and quietly moved it from one pile to another, putting his friend back into active consideration.[11]

Making the choice to leave Project Dyna-Soar, the hottest thing going in the air force, was not easy for him. "It wasn't an easy decision," Armstrong later recalled:

> I was flying the X-15 and I had the understanding or belief that if I continued, I would be the chief pilot of that project. I was also working on the Dyna-Soar, and that was still a "paper airplane," [that is, still in the planning stages] but was a possibility. Then there was this other project down at Houston, [the] Apollo program. Gemini hadn't been really much identified yet at that point. It wasn't clear to me which of those paths [would be best].[12]

Armstrong had seen many projects come and go, many airplanes and programs planned and then canceled. This complicated his choices:

> We sort of saw every project of this type as something that, it may go or it may not. Although you learn a lot when you're on a program that

eventually gets canceled, there's a lot more satisfaction in being in a program that really reaches its fruition. I can't tell you now just why in the end I made the decision I did, but I consider it as fortuitous that I happened to pick one that was a winning horse.[13]

In September 1962, Armstrong got the coveted call from the head of the astronaut office, Deke Slayton. When Slayton asked him if he still wanted to join the program, Armstrong responded with an instant yes, and within a month he was inducted into the new class of astronauts. He was the first civilian pilot to do so—when flying the X-15 he had technically been an employee of NASA.

Training for the Gemini program began immediately—there was little time to waste, and Gemini was the two-seater spacecraft that would be the bridge from Mercury to Apollo and act as a training craft for a number of Apollo astronauts, making it a choice assignment.

But before he could fly in a Gemini capsule, there was classroom preparation to be done. "NASA felt that its new astronauts with little experience with the sophistications of orbital mechanics or the differences between aircraft and spacecraft needed a quick primer," Armstrong recalled. He added, "Some of them were new to me, but overall I didn't find the academic burden to be overly difficult."[14]

Once the academic work was behind them, the astronauts hit the road, touring the new NASA facilities under construction in Florida and Texas, and visiting the many aerospace contractors that would contribute to the Gemini and Apollo programs. The astronauts were each assigned an area of specialization that they would oversee for the program—one would monitor progress in spacesuit design, another would keep an eye on flight control systems, and so forth. The schedule was demanding, with frequent flights from Florida to Houston (where the Manned Spacecraft Center was being constructed) to Huntsville, Alabama, where the Marshall Space Flight Center was building rocket boosters, and on to St. Louis, where McDonnell Aircraft was beginning work on the new Gemini capsules. Then they would travel to Southern California, home of aerospace contractors Lockheed, North American

Aviation, and the Douglas Aircraft Company. The astronauts were often on the road far more than they were at home. Armstrong loved it.

Buzz Aldrin followed a very different path into space from Armstrong or any of the other astronauts. Edwin Eugene Aldrin Jr. was born in Glen Ridge, New Jersey, on January 20, 1930. (He would later legally change his first name to "Buzz," an appellation given him by one of his two sisters, who called him "buzzer" while trying to say "brother.") Aldrin's father was an oil company executive who had served in World War One as an army pilot and gone on to become the assistant commandant of the army's test pilot school before working for Standard Oil.

Buzz Aldrin was an exceptional student, driven hard by his demanding father. Like Armstrong he was a longtime member of the Boy Scouts, and he was also on his high school football team. He was accepted to Annapolis, the naval training academy (allegedly due to his father's influence), but decided to go to West Point instead, the army's officer training academy. He enrolled in 1947 and graduated with high marks in 1951, third in his class. From there he went into the air force, training first in Florida, then at Nellis Air Force Base in Nevada.

Aldrin was assigned to a fighter squadron in 1952 and was sent to Korea, where he flew sixty-six combat missions and shot down two enemy aircraft. He was ultimately awarded two Distinguished Flying Crosses and three Air Medals for his achievements.

Upon returning to the United States, he went back to Nellis, where he worked as an aerial gunnery instructor, then was assigned briefly to the then-new United States Air Force Academy. In 1956 Aldrin was sent to Europe, where he flew patrols near the newly separated East and West Germanys. It was while serving in Europe that Aldrin first met future astronaut Ed White, who would fly in the Gemini program and perform America's first spacewalk. (White would later be killed in the Apollo 1 fire.) Aldrin returned to the United States three years later, in 1959, and enrolled at the Massachusetts Institute of Technology.

While at MIT, Aldrin saw Kennedy's address to Congress from his apartment in Cambridge, and was moved by the loud applause and apparent enthusiasm. Just a few days prior, Aldrin had received a letter from Ed White, who mentioned that he intended to apply to be an astronaut. But unlike White, Aldrin had not attended military test pilot school, a requirement for astronaut candidates. Downcast but determined, he decided to finish his PhD program at MIT before applying to NASA.

**Figure 6.6. Buzz Aldrin in air force uniform. (Courtesy of the United States Air Force.)**

"The country was swept up in the space program, and I wanted to be a part of it," Aldrin recalls. "But NASA retained its requirement that astronauts have a diploma from a military test pilot school—not one of my credentials. Since I knew that the Moon landing program Kennedy had described would need astronauts with skills other than the ones they drummed into you at test pilot school, I opted for another eighteen months of intensive work on a doctorate in astronautics, specializing in manned orbital rendezvous."[15]

He graduated with a doctorate in 1963. Immediately after graduation, and despite the test pilot requirement, Aldrin applied to NASA, but, as he suspected he would be, was turned down, so he accepted an assignment at the Los Angeles Air Force Base in Southern California, with the office responsible for planning orbital rendezvous techniques for the new Gemini program—he'd be a part of this new and exciting program one way or another. After a short time, ever determined—he'd shown that by defying his stern father's wishes throughout his school and military choices—Aldrin applied to NASA once again. As he was preparing for the new air force job, the space agency changed its rules regarding astronaut selection, and just months later Aldrin received the much-anticipated call from Deke Slayton. "We'd sure like you to become an astronaut," Slayton said. Aldrin, of course, said yes immediately. Upon arrival, Aldrin dug into his specialization—unique among the astronauts—designing orbital rendezvous techniques for the Gemini missions.

Training soon became part of the daily routine of any astronaut assigned to a mission, and it evolved quickly. As their training techniques improved, they became more targeted to the goals of Gemini and Apollo. As Armstrong explained of the training they were all undertaking, they were looking for "the best method that we could find that would give us ability to go at the earliest possible time, maximum speed, and with the highest level of confidence" to fulfill the lunar landing mandate.[16]

"I think training was about one-third of our time and effort," Armstrong said.[17] "A third had to do with planning, figuring out techniques and methods [that would get us where we needed to be to move ahead]. The last part was testing, and that's probably equal to thousands of hours in the labs and in the spacecraft and running systems tests, all kinds of stuff, seeing whether it would work and getting to know the systems very well."

Aldrin saw it similarly:

In comparison with the Gemini program, the training activities [of] the Apollo missions were quite well spelled out as to what the total mission would be. ... The training activities in each Gemini flight depended on a major degree on the preceding flight. As missions they evolved—perhaps 2–3 ahead, and they became defined working in concert with the crews—the mission plan was developed for the extravehicular activities, for the rendezvous exercises and how the experiments could be integrated.[18]

**Figure 6.7. Ed White makes America's first spacewalk during the flight of Gemini 4. (Courtesy of NASA.)**

As the training progressed, the first Gemini flights took wing. The first two missions were unmanned tests of the flight hardware, followed by Gemini 3, crewed by Gus Grissom and John Young, in March 1965. Gemini 4 followed in June, with America's first spacewalk, conducted by Aldrin's friend Ed White. Jim McDivitt was in the commander's seat. Gemini 5 flew in August, staying in orbit for a full week. Gemini 6 and 7 flew in December 1965, and represented the first close rendezvous in space—the two spacecraft closed to within just a couple of feet of each other, an unparalleled example of close maneuvering in orbit—with Gemini 6 staying in space to complete a two-week endurance mission.

Then, along with Dave Scott, it was Armstrong's turn, in Gemini 8.

The training had been as thorough as 1960s technology could make it, Armstrong recalled: "We felt it was a good representation of what we could expect, and indeed it turned out to be quite similar to what we encountered in flight. I really believed that we wouldn't have any trouble with the docking, based on the simulations we did."[19] He could not have been more wrong.

The mission of Gemini 8 was to dock with an unmanned Agena multiple times. Scott would then perform an EVA—the first since Ed White's on Gemini 4—to crawl over to the Agena and retrieve an experimental package, along with performing other EVA activities. Then Armstrong would undock from the Agena, re-rendezvous and dock, and use the Agena to change orbits. Following a number of other experiments and maneuvers, the astronauts would come home after three days in space.

That was the plan.

The Agena was an air force upper-stage rocket used to propel satellites into orbit, and had a small rocket engine that could be restarted. NASA repurposed it for the Gemini program, added a docking adapter that would fit the nose of the Gemini capsule, rendezvous beacon lights and other modifications, and dubbed it the Gemini Agena Target Vehicle (GATV). The Agena proved to be a troublesome beast in practice, however, and, as recently as the mission before Gemini 8, had malfunctioned at launch and caused a major rearrangement of flight activities. For this and other reasons, there was a bit of mistrust toward the Agena.

Despite any misgivings, the Agena target for the Gemini 8 mission launched flawlessly at 10:00 a.m. (Florida time) on March 16, 1966. Once it successfully reached a stable orbit, Gemini 8 launched just under two hours later from a nearby pad at Cape Canaveral's Launch Complex 17. All seemed to be in order as Armstrong, the mission commander, started the long process of chasing down the Agena in preparation for the rendezvous and docking. It was about 1,200 miles ahead of them when he started the hunt.

As they neared the five-hour mark of the mission, they were closing on the Agena, which they had sighted at a distance of about seventy-six miles.

As they got closer, the astronauts could make out the flashing red beacon, and their excitement increased, as reflected in their radio chatter with Mission Control.

Scott: "You're 900 feet [274 m] . . . 5 feet per second."
Armstrong: "That's just unbelievable. Unbelievable!"[20]

Then, a few minutes later:

Armstrong: "We're station keeping on the Agena at about 150 feet."

For the next half hour they maneuvered around the Agena to assess its condition, and everything checked out. Their next move would be to dock with the it and link in with the Gemini's computer control systems.

Moving at about three inches per second, Armstrong eased the nose of the Gemini capsule into the docking collar of the Agena, and the docking latches snapped closed, securing the two spacecraft firmly together.

Armstrong said, "Flight, we are docked. Yes, it's really a smoothie."

For the next twenty minutes, Armstrong and Scott verified that the Agena and the Gemini capsule were communicating properly through their connectors, while the ground controllers analyzed their own readouts, verifying what the astronauts saw. Everything looked good—the Agena was behaving perfectly. Nonetheless, Jim Lovell, another Gemini astronaut, who was acting as a CAPCOM that day, sent a gentle reminder to the crew concerning the Agena. "If you run into trouble and the attitude control system in the Agena goes wild," he said, "just send in the command 400 to turn it off and take control with the spacecraft." Command 400 would disable the Agena flight computer, and more importantly the maneuvering thrusters, and return control to Armstrong. The astronauts appreciated the nudge from Lovell, but nobody thought they would need it.

The Agena's attitude control system comprised a number of tiny rocket thrusters distributed around the fuselage. Firing these thrusters in small bursts would reorient the spacecraft. Similar systems had been in use since before the Mercury days and was considered generally reliable—but the Agena D had proved to be finicky. Command 400 was a computer code that would shut down its maneuvering system in case of an emergency.

Within minutes, the docked spacecraft slid into silence as it crossed from one ground tracking station to another—while greatly improved since the Mercury flights, NASA still had only a handful of tracking stations scattered around the Earth to communicate with the spacecraft as it orbited. What was lacking in ground control stations was augmented by a number of tracking ships at sea, however. Each of these radio dishes would hand off to the next, although there were still gaps in the system during which Gemini 8 would be out of contact with the ground.

Armstrong noted that one of those gaps happened to coincide with what would in moments become the wildest ride of his life:

We didn't have much communication with mission control. ... You see, Murphy's Law says bad things always happen at worst possible times. In this case, we were in [orbits that] didn't go over any stations. We were sort of out of radio contact most of the time, and when we were [in contact], it was [with] the ships that were at sea. They had limited ability to communicate back with mission control and transmit data to them. So our communication was just with the people on those ships, and they were trying as best they could to be helpful and identify things, but it was a real challenge for them, because there wasn't much to be gained [in the event of an emergency].[21]

After one such gap in communication, when radio contact was regained with one of the tracking ships, the scratchy radio signal carried a shocking message from Dave Scott.

"We have serious problems here," Scott said with urgency. "We're . . . we're tumbling end over end up here. We're disengaged from the Agena."

During the radio blackout all hell had broken loose. Shortly after docking with the Agena, Scott mentioned almost casually to Armstrong that they were in a slow roll—something was causing the combined Gemini and Agena to go into a spin. Armstrong tried to compensate by firing the maneuvering thrusters, but the docked pair of spacecraft continued to drift off axis.

Recalling Lovell's recommendation regarding possible malfunctions by the Agena, Scott punched command 400 into the Agena's linked computer display to shut down its maneuvering system. "We first suspected that the Agena was the culprit," Armstrong said.[22] The Gemini spacecraft had been behaving perfectly until this point in time, so he and Scott naturally assumed the trouble must be with the Agena. But the spin continued—Scott entered the Agena's shutdown command again and cycled power switches to make sure the system was operating properly. No dice.

When the rates [the spin] became quite violent, I concluded that we couldn't continue, that we had to separate [from the Agena]. I was afraid we might lose consciousness, because our spin rate had gotten pretty high, and I wanted to make sure that we got away before that happened.[23]

Armstrong said to Scott, "We're going to disengage and undock." Scott was happy to concur—getting away from the Agena stage while they still could, before the torquing forces that were rapidly building prevented them from doing so, was a good idea. The Agena was still brimming with explosive fuels, and if it ruptured it could blow them to pieces.

Armstrong released the docking latches that held the two craft together and backed away. But rather than arresting the spin, the situation rapidly became much worse. Within minutes, the Gemini capsule was tumbling along multiple axes—twisting and turning in space—at a rate that approached sixty revolutions per minute. The astronauts realized that being connected with the Agena had actually been slowing their rate of spin due to the Agena's mass. Now that they were free from it, the lighter Gemini was spinning faster. NASA engineers would later determine that a thruster control had short-circuited, and one of their Orbital Attitude and Maneuvering System (OAMS) thrusters was firing continuously, and would do so until the tanks were empty—and the astronauts dead—unless something was done, and quickly.

NASA was experiencing its first serious in-flight emergency, and communication with the spacecraft was increasingly garbled. There was nothing that could be done from the ground—it was up to the crew to save themselves. At 60 rpms, the Armstrong and Scott would soon not just be disoriented, which was dangerous, but would black out and be unable to arrest the spin. The capsule would accelerate until its fuel was exhausted, and since there is no resisting air in space, it would continue to tumble in orbit—with two dead astronauts aboard—until it eventually reentered Earth's atmosphere, still tumbling, weeks or months later.

Armstrong was trying everything he had learned in hundreds of hours of simulations to stabilize the capsule, but nothing worked. He suggested that Scott give it a try—Armstrong thought he might be missing something, but despite Scott's best efforts he too was also unable to stabilize the capsule.

Armstrong told Mission Control in a strained voice, "We're rolling up and we can't turn anything off. Continuously increasing in a left-hand roll."

The CAPCOM said, "Roger." There was little else to add. The flight controllers waited helplessly. Then, after a few more sweaty minutes, a crackly Dave Scott came back on the radio: "Okay, we're regaining control of the spacecraft slowly, in RCS direct."

Armstrong had used the only option left open to him. With the OAMS system clearly malfunctioning, he had turned it off, and fired up the Reentry Control System, or RCS. This was a second, completely separate set of thrusters, powered by much smaller fuel tanks, designed to be used only during reentry to keep the capsule oriented properly and prevent it from burning up or overshooting the splashdown zone.

With the RCS system activated, the mission rules stipulated that they reenter immediately—once the RCS was started, you could not be certain that the thrusters would not leak fuel, and if they leaked too much it could jeopardize their ability to hold their trajectory during reentry.

The flight director on duty, John Hodge, declared that the mission was at an end, and that the astronauts should prepare for reentry. But since they would not be reentering on schedule or on target, Mission Control had to compute a new landing zone in the Pacific Ocean. They would ultimately splash down near Okinawa, Japan.

Armstrong, now remarkably calm after their brush with infinity, said to Scott, "Okinawa. Well, I'd like to argue with them, about the going home, but I don't know how we can." Scott agreed.

As it was, the navy would have to scramble to send ships to pick them up within a reasonable time, and the astronauts could be drifting in tropical seas for hours, cooped up inside a bobbing, heaving capsule. Gemini was a sturdy spacecraft but a lousy boat, as Armstrong later commented with a wry grin.

Soon after splashdown, a navy plane dropped divers near the capsule to attach floats that would prevent it from sinking even if it began to take water. But that did not stop the bobbing. Both astronauts became seasick, and the small supply of vomit bags was used up quickly.

Two hours later, changed out of their sweat-soaked and vomit-streaked spacesuits, the astronauts were aboard a navy vessel headed for Hawaii. Scott later said of Armstrong: "The guy was brilliant. He knew the system so well that he found the solution. He activated the solution under extreme circumstances. . . . It was my lucky day to be flying with [him]."[24]

**Figure 6.8. Armstrong and Scott, seasick and queasy from being cooped up in the bobbing capsule, await recovery by the navy. (Courtesy of NASA.)**

Armstrong and Scott headed for their debriefings, reunions with their families, and then back intro training for their next missions on Apollo.

Buzz Aldrin had his Gemini flight just nine months later, on Gemini 12. This was the last Gemini flight, and the problems of working in space, during an EVA, had still not been ironed out. This was considered critical for the Apollo flights, and Gemini 12 would be the last opportunity to get it right.

And it was not for a lack of effort. Since Ed White's spacewalk in June 1965, three more Gemini flights had attempted to perform useful tasks in EVA. Gene Cernan gave it his best on Gemini 9 but ended up overexerting himself, and by the time he got back into the capsule he was wheezing and had sweated so much that his visor was completely fogged over. The only way he could see well enough to return to the cockpit was to rub a circle with the tip of his nose and sight through the tiny clear spot. His pulse rate was over 180 beats per minute at its peak—dangerous enough on the ground; more worrisome in space. He had been at it for just over two hours when he called it quits, having achieved little.

On Gemini 10, future Apollo 11 astronaut Mike Collins performed two EVAs. The first was a more cautious approach, called a "stand-up EVA," in which he opened the hatch, stood on his seat and performed photographic experiments. Later in the flight, after they had chased down and rendezvoused with (but not docked with) the derelict Gemini 8 Agena, he performed a second, far more ambitious spacewalk. Using a hand-held gas-jet maneuvering gun (similar to one used by White on Gemini 4), Collins maneuvered from the Gemini to the Agena to retrieve a micrometeorite experiment. By the time he got back into the cockpit, he too was exhausted, his frustration heightened by the lack of handholds on the Agena to assist him—it was mostly smooth and terribly difficult to crawl across without drifting away. This second attempt had lasted about forty minutes.

Gemini 11, the penultimate flight of the program, saw Dick Gordon performing two EVAs. During the first spacewalk, Gordon's primary objective was to attach a tether to the docked Agena for a "passive stabilization" experiment, in which the two spacecraft would separate and see if one stabilized the other. The EVA was scheduled to last two hours but was terminated after only about thirty minutes due to that now-familiar enemy, exhaustion of the spacewalker. The second EVA was a "standup" EVA, and Gordon photographed stars and the Earth and lasted two hours. But doing useful work in zero-g still eluded Gemini.

Back on Earth and in training for Gemini 12, Aldrin was tracking the results of these efforts carefully. He was a strong proponent of additional training, which some of the others had eschewed. Aldrin did countless flights in NASA's zero-g training plane, dubbed the "vomit comet," which was a modified passenger jet that flew parabolic arcs up, then down, then up, then down again. Each time it dove, the astronaut in the passenger area, which was empty save for a simulator with handholds, got a limited amount of time in a weightless state. But these were very brief periods, which Aldrin knew would not be sufficient to master the chores assigned to him.

For more extended EVA practice, Aldrin went over to a local high school, where NASA had arranged to rent time in the pool. (They had to schedule around water polo practice—the agency did not yet have its own water-training facility.) Aldrin spent hours in a Gemini-Agena simulator that NASA left submerged in the deep end of the pool. It was modeled to simulate the newly designed handholds on the Agena, as well as the back area of the Gemini capsule, where a "busybox" was installed. This was a small container with various manual activities designed for an astronaut to demonstrate manual dexterity in zero-g. Into the water he went, wearing a Gemini spacesuit modified for use in the pool, and there he would spend hours practicing each move again and again, until he could practically do them in his sleep. Some of his comrades thought he was being excessive, but this was Aldrin's way—study a problem, work up a set of possible solutions, and practice them until he was certain that he had found the proper solution. It was the same way with the other aspects of the flight; the navigation of the spacecraft in orbit—along orbital paths that he had helped to design due to his experience at MIT—was practiced on paper with the use of star charts until he was certain he know it inside and out.

His practice would pay off, and handsomely.

On November 11, the target Agena launched just after 2:00 p.m. Eastern time, followed by the launch of the crew in Gemini 12 at 3:46 p.m. The crew—James Lovell as mission commander and Aldrin as pilot—began chasing down the Agena for their first docking. This involved laboriously entering data into the Gemini flight computer, a painstaking process. While the basic programs were hardwired into the system, any changeable parameters had to be entered via endless keystrokes on a small numeric pad.

A half hour later, the Gemini's radar acquired the Agena on the first try. "Houston, be advised, we have a solid lock-on, 255.5 nautical miles [473 km]," Aldrin radioed down.

In the midst of this, however, their radar stopped working, and they were unable to maintain a fix on the Agena, which could easily have scrubbed this part of the mission. But Aldrin was prepared: "The fallback for this situation was for the crew to consult intricate rendezvous charts—which I had helped develop—to interpret the data using the 'Mark One Cranium Computer' (the human brain), and then verify all this with the spacecraft computer," Aldrin would later say.[25] He reached under his seat to retrieve a sextant. Using this, a slide rule, and paper and pencil, he calculated where he thought the Agena was and the trajectory needed to reach it.

This manual navigation exercise was something that had been scheduled for later, and was this only the second time it had been attempted in flight. The prior effort had been less than satisfactory, but manual navigation was another one of the items on the Apollo-preparation checklist. Nobody wanted to be stuck in lunar orbit in the Lunar Module, unable to find the Command Module—their ride home—due to a failed computer or radar.

Based on Aldrin's calculations, they not only managed to find the Agena and dock, but did so using less fuel than any previous flight.

Lovell practiced docking and undocking a few times. The flight plan called for using the Agena to boost them into a higher orbit. This activity was canceled, however, due to the fact that the Agena indicated possible issues with its rocket engine, and the flight director decided to abort the exercise. Disappointed, Aldrin and Lovell settled in for a meal and rest period.

Twenty-one hours after launch, Aldrin performed his first space-walk—another stand-up EVA, rising out of the Gemini hatch but not leaving the spacecraft. Like others before him, Aldrin performed photographic experiments and retrieved a micrometeorite experiment situated near the hatch. Then he had a small surprise.

"During the second night EVA pass I saw blue sparks jump between the fingertips of my gloves," he later said. "Space clearly was not just an empty void. It was full of invisible energy: magnetism and silent rivers of gravity. Space had a hidden *fabric*, and the fingers of my pressure gloves were snagging the delicate threads."[26] He completed his assigned tasks and reentered the spacecraft for a long rest period.

The next EVA would be the dealmaker. Or breaker. He would attempt, in this final flight of the program, to prove that an astronaut could accomplish routine chores in zero-g. A lot was riding on this—not just the future of Apollo but proof that Aldrin's many months of intensive and self-motivated underwater training was a valid way to approach the manual dexterity in weightlessness required for Apollo.

About forty-three hours into the mission, Aldrin hooked up to an extra-long umbilical and exited the depressurized Gemini capsule, with Lovell feeding out the hose as Aldrin moved away from the spacecraft. Carefully, using new handholds installed for this purpose, Aldrin made his way across the nose of the capsule and onto the Agena. Working his way slowly forward, hand over hand, he prepared a "gravity gradient" experiment for later, easily went back to the capsule to exchange cameras with Lovell, then moved to the back end of the Gemini space-craft, which had a hollow adapter ring to which the "busybox" was affixed. He showed little exertion throughout, possibly irritating some astronauts who had gone before him.

When he reached the rear area of the Gemini capsule, Aldrin slipped his boots into newly designed foot restraints that would help keep him properly oriented in the frictionless environment of space. In earlier attempts, despite some level of preparation and special tools, the other astronauts had found themselves twisting and torquing when they tried to use tools to loosen or tighten nuts and bolts. But, by using the experience gained in his underwater simulations and his sessions in the "vomit comet," Aldrin executed the assigned tasks with careful precision, minimizing motion and torque.

For the rest of the EVA, he checked off each step in his mind as he went through the long list of tasks. He then moved back to the hatch of the capsule, cleaned the windows, and went inside after two hours in open space.

While the experience gained during each EVA attempt had contributed to the database of knowledge on the rigors of working in space, it was not until Gemini 12, with arduous preparation and forethought, that the program got it right. And just in time.

After a couple more days, and one more stand-up EVA, Lovell and Aldrin entered the reentry codes into the computer and headed to a splashdown near Bermuda in the Atlantic Ocean.

As Aldrin later reflected, "Project Gemini had finally triumphed. *All* of its objectives had now been met. We were ready to move on to Project Apollo and the conquest of the moon."[27]

It was time to prepare for Apollo, and the rigors of the long missions to the moon. Back into the simulators the astronauts went, assigned either as flight crews or backups to those crews.

Then came the Apollo 1 fire (see chapter four). The loss of three fellow astronauts, especially on the ground as opposed to during flight (which was perceived as being far more dangerous), made it all the more difficult to accept. "I remember it very well," Armstrong recalled. "I'd known Gus [Grissom] for a long time." Armstrong was close friends with Ed White as well. "You know, I suppose you're much more likely to accept loss of a friend in flight, but it really hurt to lose them in a ground test. That was an indictment of ourselves. I mean, [it happened] because we didn't do the right thing somehow. That's doubly, doubly traumatic."[28]

Aldrin recalled the scene as laid out by the crew's would-be rescuers: "The technicians saw that the entire cabin had been swept by flames. Ed White and Gus Grissom lay in a jumble on the cabin floor just below the hatch frame. Fire had fused the material of their suits with the melted plastic and metal near the hatch. Roger Chaffee still lay on his couch, his restraining straps and oxygen hoses disconnected. All three were dead."[29]

Upon hearing the news, Aldrin could only think of his friend Ed White, remembering a day years before when the two of them had first discussed the possibility of becoming astronauts: "He had been so confident then, tall and husky, what everyone thought of as an all-American fighter pilot. Now he was gone. So was Gus. So was Roger."[30]

Horrifying as the Apollo 1 accident was, however, there were positive outcomes. NASA oversaw the design of a much safer spacecraft, and got a much firmer grasp on the contractor, North American Aviation, who was building it—they had been lax in safety and quality control. In addition, "We were given the gift of time," Armstrong said. "We didn't want that gift, but we were given months and months to not only fix the spacecraft, but rethink all our previous decisions, plans, and strategies, and change a lot of things, hopefully for the better."[31]

While NASA investigated the fire, training for Apollo continued, and the crews threw themselves into the routine as somewhat of a balm for the incandescent trauma of Apollo 1.

Armstrong recalls the training for Apollo as more focused than previous efforts:

> It was very goal-oriented. We tried to define it as narrowly as we [could], rather than as broadly as you would in research, because with the time constraints that we were facing then, the desire to get there as fast as we can, we were in a race and that was very evident to us all the time. You wanted to not be diverting your attention in any way to things that you really did not need to worry about. You wanted to focus on all the things that you knew you had to do and had to master. That was the principal difference as we went into the Apollo flights.[32]

Notably, the equipment was improved as well. Previously, computers had been used for calculations—what trajectories were involved, and specific outcomes that were desired. These were used as parameters to be entered into the analog electromechanical machines that drove the simulators, which had their roots in WWII trainers. As computer technology advanced in the 1960s, due in no small part to the needs of the space race, digital machines played an increasingly central role in making simulations much higher fidelity and more responsive.

"In the late '60s our computer simulations were really quite excellent," Armstrong said. "They were quite adequate to do most all the things that we were doing. There's an old perception that simulators are always more difficult to fly than the craft themselves. In general, that is true, and it's certainly turned out to be true in Apollo, particularly the lunar module, which was to our benefit that it was easier to fly than the simulator, because we were expecting something that was somewhat more cantankerous and contrary than it actually turned out to be."[33]

As the first flight of Apollo drew closer, the simulations intensified. All the bugs had to be worked out before a flight of the new hardware could be attempted, and the simulations now involved a spacecraft simulator for the crew on one end, a fully staffed Mission Control in Houston on the other end, and banks of intermediary computers in between that ran the simulations. The computers were overseen by the simulation supervisors (SimSups, see chapter 4) who devised the problems to be solved and added complications, often devilishly difficult ones, that both parties would have to work through.

But not everything could be simulated via computers. Flying the Lunar Module, and landing it on the moon, would require skills impossible to gain in only in computerized cockpits, and for this, a special flying machine was devised. It was called the Lunar Landing Research Vehicle (LLRV), and later the Lunar Landing Training Vehicle (LLTV), but the astronauts referred to it as the "flying bedstead." And it looked like one—a spindly-legged platform with a small gondola on the front, wrapped around a powerful jet engine that pointed downward. Small gas thrusters were affixed to each axis. This ingenious machine used the jet engine to the machine into the sky, and the gas thrusters were used to maneuver. It would ascend to the desired altitude, then the jet engine would be throttled down to a hover—to simulate five-sixths of Earth gravity, that of the moon—and the thrusters took over to maneuver in altitude and sideways motion. It was a frighteningly difficult machine to fly, and NASA thought it so dangerous that they more than once considered discontinuing its use—they did not want to lose any more astronauts.

The astronauts who trained on the LLTV approached the machine with both caution and respect. In Armstrong's view, however, it was indispensable, and, despite the perceived dangers, he lobbied for its continued use.

The LLTV was actually designed long before Apollo training began, and was in fact in the pipeline before the Lunar Module's design was finalized in the early 1960s.

Armstrong spoke later of the LLTV's origins:

We decided that [Apollo] was going to be a pretty complicated project, and what we should do . . . was build a little device, a little one-man device which would just investigate the qualities and requirements of flying in a lunar [environment]—to build the database from which we would build the bigger thing to carry the real spacecraft.

So we actually devised such a craft. It looked like a tin Campbell Soup can sitting on top of some legs, with a gimbaled [jet] engine underneath it. That became the basis for what went out as a require-ment for bid to build the LLRV, Lunar Landing Research Vehicle. It was not known at this time that there would be a lunar module. . . . Matter of fact, the lunar module came after the lunar landing research vehicle. Fortunately, the characteristics and the size [and] in the inertias and so on of this training device were very much like the lunar module. That was strictly fortuitous.[34]

Once the LM's characteristics were known, the Lunar Landing Research Vehicle was redesigned and debuted as the Lunar Landing Training Vehicle. Armstrong said that the LLTV "was designed to be even more LM-like, so it would give you a good representation. In fact, it did. All the pilots, I think, to my knowledge, . . . thought it was an extremely important part of their preparation for the lunar landing attempt."[35]

## LLRV PROFILE VIEW

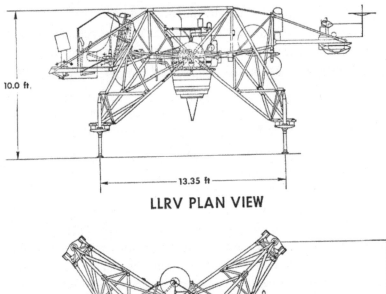

10.0 ft.

13.35 ft

## LLRV PLAN VIEW

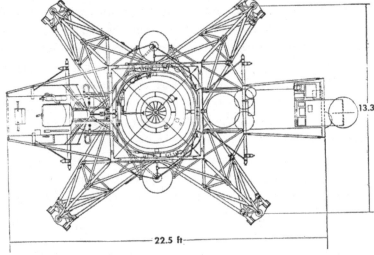

13.3

22.5 ft

Figure 6.9. A diagram of the Lunar Landing Research Vehicle, which was nearly identical in appearance to the later Lunar Landing Training Vehicle. The jet engine is at center, and the pilot gondola to the left. (Courtesy of NASA.)

He added,

> It was harder to fly than the lunar module, more complicated, and subject to the problems that wind and gusts and turbulence and so on introduce, that you don't have on the moon. The systems were somewhat choppier or less smooth than the actual Lunar Module, both propulsion and attitude control systems were so. The lunar module was a pleasant surprise.[36]

Training on the LLRV/LLTV was indeed treacherous, with the jet engine howling directly behind the pilot, and the craft in a continual state of trying to slide off to one side and topple, with potentially explosive results. It was like balancing a bowling ball on a broomstick, with a high and unstable center of gravity that shifted as the fuels were consumed at alarming rates. Gusts of wind could push it to one side, and instability was assured. All these factors caught up with Armstrong during a training run on May 6, 1968, at Edwards Air Force Base. He was flying at about 100 feet (the machine's maximum altitude was limited to about 500 feet), and as he was maneuvering the LLRV developed a control problem—the thrusters were suddenly starved of fuel, and gusts of wind popped up. The craft started to dip on one side, then, having lost its balance, to slide toward the ground. Armstrong struggled to regain control, but he had only seconds, as he was not high enough to have much time before it crashed. At the very last moment, he fired the ejection seat, which was powered by small rockets, and boosted just high enough for his parachute to open before slamming into the desert floor.

As Armstrong later put it, in typically understated engineering prose, "There was very little time to analyze alternatives at that point. It was just because I was so close to the ground, below 100 feet in altitude. So again, time when you make a quick decision. You departed."[37] Indeed he did, in a fiery and furious fashion. The LLRV crashed seconds later, igniting a huge fireball, and Armstrong came down not far away. He was shaken but unhurt, and when he returned to the base he did not even bother to mention the occurrence to his colleagues. It was just another day in the life of a test pilot to him.

Five of these bizarre craft were built, with three being destroyed in crashes—not a promising record for an aircraft of any kind. Fortunately, the actual Lunar Module was a much superior flying machine.

As training proceeded, Apollo grew wings. On October 11, 1968, Apollo 7 lifted off from the Cape on the shakedown cruise that Apollo 1 had been intended to be. Aboard were Wally Schirra, Walt Cunningham, and Donn Eisele. This was an Earth-orbital flight only, but it was nonetheless daring, with ten days spent in orbit aboard a new and untried spacecraft. The Apollo Command Module passed with flying colors.

There were a number of incremental development flights planned between Apollo 7 and any lunar missions, but with the Lunar Module behind schedule (it was too heavy to accomplish its mission, and drastic weight-reduction efforts were underway), and Kennedy's end-of-decade deadline looming, something had to give. With a speed unimaginable today, NASA reshuffled the flight schedule, and just three months after Apollo's inaugural flight, Apollo 8 was sent to orbit the moon. There was no Lunar Module on this flight, and, had anything gone wrong with the spacecraft's main engine, the three astronauts—Frank Borman, Jim Lovell, and Bill Anders—could have been permanently stranded in lunar orbit. It was an audacious plan, but fortunately everything worked perfectly.

Armstrong was as surprised as any of the other astronauts when the announcement was made about the decision to go to the moon on only the second flight of Apollo—and the first manned flight of the Saturn V moon rocket. The Saturn V was experiencing its own developmental problems, not the least of which was that its main rocket engines were pulsing violently during ascent, a condition they called "pogo."

Armstrong thought the plan for Apollo 8 to orbit the moon to be "very bold"

> because we still had the Pogo problem on the Saturn and we'd had a couple of problems with [prior unmanned] Saturn V launches, so to take the next one, and without those problems being demonstrated as solved, and put men, a crew on it, not just take it into orbit, to take

it to the Moon, it seemed incredibly aggressive. But we were for it. We thought that was a wonderful opportunity. If we could make it work, why, it would make us a giant jump ahead. . . . It was kind of a complex process, but it showed a lot of courage on the part of NASA management to make that step.[38]

Armstrong and Aldrin were on the backup crew for Apollo 8. Mike Collins, the Command Module Pilot had been as well, but when he experienced some medical difficulties he was replaced. On December 21, 1968, the trio watched as the crew they were backing up launched for their six-day trip to the moon and back. It's never easy to watch astronauts fly a mission you also prepared for, but any disappointment they might have felt was short lived.

Armstrong was called in for a meeting with Deke Slayton as Apollo 8 circled the moon. "During the flight of Apollo 8 I had three or four meetings with Deke Slayton about, first, would I take the third one down," referring to the third lunar mission, "and then we had a lot of talks about who might be available and be right to be on that crew, that sort of thing."[39]

After Apollo 8, Apollo 9 launched on March 3, 1969, for another Earth orbital mission, this time with a Lunar Module, the first time it was flown by a crew. For ten days, the crew—Jim McDivitt, Dave Scott, and Rusty Schweickart—put the combined system through its paces, thoroughly testing every phase of docking, undocking, firing both the LM's descent and ascent engines, and maneuvering. Everything went as planned, other than a bit of space sickness on Schweickart's part.

Apollo 10 was the last test before the lunar landing, and it launched on May 18. It was only the second flight to the moon, and the first to carry an LM there. Over the course of eight days, everything was practiced except for the actual touchdown—Tom Stafford and Gene Cernan flew the LM down to within a few miles of the lunar surface, then dropped the descent stage and flew back up to orbit to rendezvous with John Young, who was waiting for them in the Command Module. The mission went well, with the exception of a few moments when the LM was staging—the ascent stage (the cabin with the astronauts inside) spun wildly immediately after separating from the descent stage before the astronauts could regain control. But this turned out to be a procedural problem, not a hardware issue, and did not jeopardize the next mission—Apollo 11.

Training for Apollo 11 continued at a frenzied pace, with the launch just four months away. "Michael Collins spent all of his time mastering the command module," Armstrong recalled. "Buzz Aldrin and myself focused a great deal of our time on mastering the Lunar Module, knowing it inside out, and then we had, of course, to learn the experiments and the lunar surface science work and the installation of equipment on the Moon, and all that kind of stuff. It took a lot of time. I suppose we would have liked a little more time, but when the time came, we had to say, yes, we were ready to go."[40]

Figure 6.10. A preflight press conference for Apollo 11. From left, Aldrin, Armstrong, and Collins. (Courtesy of NASA.)

At a press conference eleven days before launch, Armstrong, Aldrin, and Collins responded laconically to last-minute questions from a corps of reporters. Some of the questions were rather trivial— for example, asking how the Command Module came to be named *Columbia*. "Columbia is a national symbol," Armstrong said, "and, as you all know, it was the name of Jules Verne's spacecraft that went to the Moon."[41]

Then a reporter bore in, and asked Armstrong, "Have [you] decided on something suitably historical and memorable to say when you perform this symbolic act of stepping down on the Moon for the first time?"[42]

Armstrong was careful in his answer. The media had speculated long and hard over what the first words from the first man taking the first step onto another world might be. Whatever they were, they would reflect the sweat and treasure of a nation for the better part of a decade. There was a lot at stake, and insofar as anyone knew, a NASA public relations team had been busy creating the perfect words to be spoken at that historic moment.

As it turned out, however, this was not the case. Rather surprisingly, given NASA's control-oriented culture, this decision had been left up to Armstrong. As the issue of what to say on the moon had been batted around NASA, the head of public relations, Julian Scheer, had written a memo that stated, in effect, that since Queen Isabella had not specifi- cally instructed Christopher Columbus on what he would say when he encountered land during his journey to the New World, NASA would not be telling Armstrong what to say when he stepped onto the moon. It had become a sensitive point, though you would not guess it from Arm- strong's response to the reporter: "No, I haven't." And that was that.

Armstrong later said,

The late Julian Scheer, who really led the NASA relations with the outside world in many ways, was absolutely adamant that headquar- ters never put words in the mouths of their people, not just astro- nauts, but anybody, that they let people speak for themselves. They made it known sort of what the party line was and what the NASA position was, but beyond that, they never, to my knowledge, controlled the . . . public statements of others. Certainly they insisted, in the case of the flight crews, that they not be told what to say, that their statements be their own elocution of what they saw and what they wanted to say. As far as I know, that prohibition was never violated.[43]

And what of those famous words Armstrong selected? "It was not something that I really concentrated on but just something that was kind of passing around subliminally or in the background. But it, you know, was a pretty simple statement, talking about stepping off something. . . . It wasn't a very complex thing. It was what it was."[44]

Finally, a reporter asked Armstrong, "What would, according to you, be the most dangerous phase of the flight of Apollo 11?" Armstrong said, "Well, as in any flight, the things that give one most concern are those which have not been done previously, things that are new." He would not be drawn into any attempts at melodrama. "The LM engine must operate to accelerate us from the Moon's surface into lunar orbit, and the Service Module engine, of course, must operate again to return us to Earth." He added, "As we go farther and farther into spaceflight, there will be more and more of the single-point systems that must operate." He closed with, "We have a very high confidence level in those systems, incidentally." The engineer had spoken. End of story.[45]

But, in fact, Armstrong had harbored some concerns about parts of the hardware, in particular the LM's ascent engine. It was a simple device, engineered by aerospace contractor Bell Aerosystems. The engine was fueled by hypergolic fluids, highly reactive (and toxic) chemicals that explode on contact—no igniter needed. The engine was pressure fed, with the fuels pushed out of their storage tanks by inert gas. In short, all you had to do was open the valves and *wham* off you went into the sky. Armstrong's concerns had centered on those two valves. Like much of 1960s rocketry (and rocketry to this day), many of the mechanical functions were handled by small explosives. All that was required for the ascent engine to operate was an electrical jolt to those tiny explosive squibs, and when they exploded, they would force the fuel valves open. But what if the electrical signal shorted out, Armstrong wondered, or what if the explosive squibs didn't fire? He inquired as to the possibility of mechanical overrides for the valves—a couple of manually activated handles—being installed on the engine. The answer that came back was no, it would add complexity and weight. The engineers at Bell Aerosystems assured NASA that the system was perfectly reliable. In the end, Armstrong reluctantly agreed.

On the morning of July 16, 1969, Armstrong, Aldrin, and Collins were awakened at 4:00 a.m. by chief astronaut and compatriot Deke Slayton. They were sleeping in the crew quarters at the Kennedy Space Center and had been effectively quarantined for the better part of a week before the flight to prevent any kind of airborne illness striking a crewmember on the way to the first moon landing. After routine pre-flight medical checks, they ate the traditional preflight astronaut fare: steak, eggs, toast, fruit juice, and coffee.

Apollo 8 astronaut Bill Anders joined Slayton and the crew at breakfast. Then, about a half hour later, the astronauts went upstairs to suit up in preparation for launch, just a few hours away.

A NASA van soon drove them to the launchpad, eight miles away. The rocket stood before them, magnificent in the dawn light, shrouded with vapor trails from the cryogenic fuel boiloff.

As the astronauts rode the elevator to the Command Module, some 360 feet (110 m) above the ground, astronaut Fred Haise was completing a 417-step checklist inside the capsule. Every switch had to be in the proper position, and every gauge and indicator light had to show the proper reading.

The astronauts were helped into their assigned couches by the pad team, their shoulders touching inside the capsule due to the bulky suits. Once en route to the moon, and in the zero-g environment, the capsule would seem much larger—especially once they had shed the pressure suits.

Inside the Command Module, as the count neared the final minute, Armstrong calmly moved his hand to the abort handle, a T-shaped switch mounted on the arm of his couch. In the event that anything went wrong during launch, all he had to do was twist the handle and the emergency escape system—a set of rockets that sat atop the Command Module—would fire and blast them away from a malfunctioning booster. It was a last resort, and not something they liked to think about—it had never been tested with humans. But it was a necessary precaution, and just a couple of nights before, NASA administrator Tom Paine had told Armstrong, "If you have to abort, I'll see that you fly the next Moon landing flight. Just don't get killed." It was his way of saying, "Don't wait too long to abort if things go south on you."[46]

At about the same time, Aldrin, who was in the center seat, turned first to Armstrong, then to Collins, with a wide grin. After years of preparation and effort, they were actually going to the moon.

Then, at T-minus-eight seconds, all hell broke loose.

Flames leaped toward the deflection trough below the launchpad and billowed out to the sides. The fuel-heavy Saturn was shrouded in smoke and steam as it burned through 23 tons of kerosene and liquid oxygen even before leaving the ground. The power of the five huge rocket engines, each producing 1.5 million pounds of thrust, built up, with the rocket shaking as it was held in place by the hold-down clamps.

The launch commentator counted down the last seconds, then at T-minus-zero, said: "Lift-off! We have a lift-off, thirty-two minutes past the hour. Lift-off on Apollo 11."[47]

Armstrong said, "We've got a roll program," as the rocket rolled onto its side to head off at an angle toward the equator. He sounded like he was giving someone the time of day—almost laconic.

A few minutes later the first stage separated explosively from the second stage and fell to the Atlantic Ocean, far below. The second stage's five J-2 engines lit up, continuing to drive Apollo 11 into orbit. The rocket continued on into orbit, with three grinning astronauts on their way into history.

Nine minutes after launch, the S-II's engines were shut down, and the Saturn V staged again. The single J-2 rocket engine on the S-IVB stage ignited, burning for a shade over two more minutes and placing Apollo 11 into the proper orbital trajectory. At eleven minutes and forty-two seconds into the flight, the third stage shut down, and the spacecraft coasted into orbit.

"Shutdown," Armstrong relayed. Mission Control gave them a go for continuing to prepare for one more firing of the S-IVB's rocket engine to leave Earth orbit and head to the moon. Two hours and forty-four minutes later, after a complete check of all their spacecraft systems, the single rocket engine on the third stage fired for the last time, sending them into translunar injection (TLI)—they were off to the moon.

After thirty more minutes, Collins took control of the Command Module and separated from the third stage. He pulled ahead of it, turned 180 degrees, then slowly boosted back toward the S-IVB. Nestled inside it, atop the fuel tanks and rocket engine, was the Lunar Module—four side panels had been released and the LM was crouched atop the stage, its legs folded, waiting to be plucked free.

Collins homed in using ranging radar and an optical sight—this maneuver was as much an art as it was about technology. He carefully slid the docking probe into the receiver atop the LM and closed the docking latches. They would not be released until his crewmates headed down to the moon.

"That wasn't the smoothest docking I've ever done," Collins said after the maneuver was completed—he felt that he had done better in many of the countless simulated docking runs. But Armstrong said, "Well, it felt good from here." Within ten minutes they were pressurizing the LM, and just under an hour later Collins pulled the LM free of the third stage.

As the two spacecraft coasted to the moon, Armstrong radioed to Mission Control what he was seeing out the window:

We didn't have much time, Houston, to talk to you about our views out the window when we were preparing for LM ejection; but up to that time, we had the entire northern part of the lighted hemisphere visible including North America, the North Atlantic, and Europe and Northern Africa. We could see that the weather was good all—just about everywhere. There was one cyclonic depression in Northern

Canada, in the Athabasca—probably east of Athabasca area. Greenland was clear, and it appeared to be we were seeing just the icecap in Greenland. All North Atlantic was pretty good, and Europe and Northern Africa seemed to be clear. Most of the United States was clear. There was a low—looked like a front stretching from the center of the country up across north of the Great Lakes and into Newfoundland.[48]

**Figure 6.11. The Command Module and Lunar Module are seen docked, as they looked in transit to the moon. (nerthuz © 123RF.com.)**

They were now able to get out of the pressure suits they had been wearing since hours before launch and relax for a bit.

Sixty-one hours into the mission, Apollo 11 passed an invisible point in space called the *equigravisphere*, where the gravitational pull of the Earth and moon are equal. From here on out, they would accelerate toward the moon as it tugged at their spacecraft. Hours later Collins fired the single large engine at the back of the Command/Service Module (CSM), which was now pointed toward the moon. This engine burn would position them to swing into an orbit of about sixty miles above the lunar surface. One more firing would be done on the far side of the moon, out of radio contact with Earth, to lock them into lunar orbit.

As they passed across the back side of the moon, Aldrin marveled at the raw nature of its surface, which is much rougher on the far side—the one we never see from Earth—than on the near side. The moon is "tidally locked" with the Earth, meaning that one side forever faces our planet.

"The back side of the Moon was much more rugged than the face we saw from Earth," Aldrin later said, "This side had been bombarded by meteors since the beginning of the solar system millions of centuries ago."[49]

When the engine burn was complete, Collins remarked, "Well, I don't know if we're 60 miles or not, but at least we haven't hit that mother." Aldrin read off the altitude indicator: "Look at that! Look at that! 169.6 by 60.9!" Collins replied, "Beautiful, beautiful, beautiful, beautiful!"[50]

About ninety-five hours after launch, Armstrong and Aldrin suited up and entered the Lunar Module to prepare for the lunar landing.

The two astronauts completed their checklists and reported to Mission Control that everything was ready to go. In the meantime, Collins worried from his station in the Command Module *Columbia*. Though he had not said it out loud, he gave his companions a roughly fifty-fifty chance of success. That was what he was *thinking*—what he said was, "You cats take it easy on the lunar surface. If I hear you huffing and puffing, I'm going to start bitching at you."[51]

Collins hit the switch that released the *Eagle.* The docking mechanism had a spring that gently pushed them apart. The force of that spring, along with every other maneuver, no matter how small, had all been factored into their trajectory calculations for the landing. What had *not* been factored in was that there was still some excess air in the tunnel between the two spacecraft, and when they separated there was a slight "pop," like a giant champagne cork. None of them noticed it, but the added push was enough to affect the speed of the LM—something that Armstrong would only notice as the LM got closer to the lunar surface.

Armstrong fired the LM's thrusters to maneuver away from *Columbia*, then slowly rotated the LM in front of Collins so that he could do a visual scan of the lunar lander. Collins saw no problems—there was no damage to the LM, and more importantly all four landing legs were extended and locked (they had been folded inward for launch)

"I think you've got a fine looking flying machine there, *Eagle*, despite the fact you're upside down," Collins said.[52]

Armstrong replied, "*Somebody's* upside down."

After a thoughtful pause, Collins said, "You guys take care."

Armstrong replied, simply, "See you later," as if he were heading out to for just another day at the office.

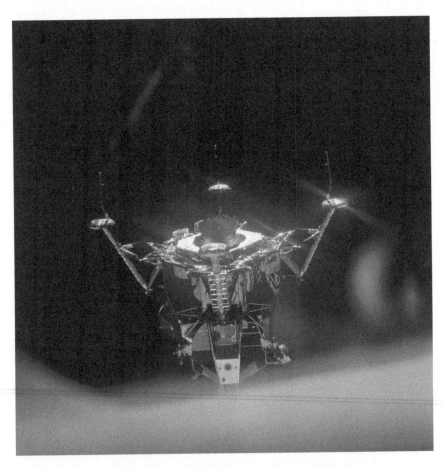

**Figure 6.12. The Lunar Module *Eagle* just prior to lunar descent. (Courtesy of NASA.)**

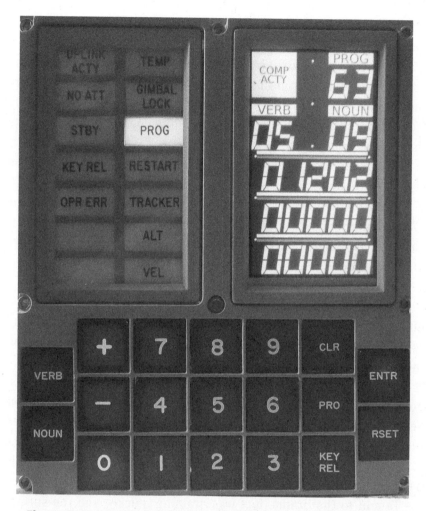

**Figure 6.13. A simulated image of the 1201 and 1202 alarms on the Lunar Module's computer. (Courtesy of Nick Howes.)**

*Eagle* continued to orbit for another two hours and twenty minutes as Armstrong and Aldrin finished preparations to head down to the surface below. They were soon cleared to begin their descent, and, as they fired the LM's rocket engine to slow them down and allow them to begin their fall to the targeted landing zone, Armstrong noticed that the landmarks were not lining up properly—they appeared to be about two seconds ahead of when they should have appeared against the numeric indicators scribed on the window. There would be some extra work to find a proper landing site as they closed on the rocky surface.

Armstrong and Aldrin were standing up in the LM's cabin for the landing—early iterations of the LM had included seats, but they were removed due to weight considerations. They were just passing 33,000 feet when a shrill alarm sounded—it was the computer alerting them that it was having a problem.

Armstrong said, "Program alarm," and then, "It's a 1202." Aldrin repeated it to make sure the controllers in Houston heard it—"1202."[53]

A few moments later, Armstrong said, "Give us a reading on the 1202 program alarm," to Mission Control. There was no panic in his voice, no real sign that he was concerned—it was just a bit more urgent in tone.

Mission Control told them that they were still clear to land.

As *Eagle* continued its descent, the computer continued to lock up intermittently and would stop displaying their range to the surface and speed. But it would then come back online, and with Aldrin reading off the numbers as Armstrong closed on the landing zone.

Armstrong had already realized he was way off target, but just how forbidding the terrain below would be had been unknown. It was becoming increasingly rugged, and his attention was intently focused through the LM's tiny triangular window as he searched for a flat place to set down. Aldrin was now manning the computer, reciting a steady narrative of altitude and speed.

To add to the drama, fuel was draining quickly, and the surface scooting below them continued to be rough and dangerous. He slowed their decent to a hover and scanned ahead, seeking anything that looked flat and smooth. Any surface irregularity more than a couple of feet tall or deep could cripple the LM, or at least make it a challenge to depart when it was time to return to orbit.

Tense moments later, Aldrin said, "I got the shadow out there," then, looking back at the computer readout, "250 down at 2½. ... Coming down nicely."[54]

"Going to be right over that crater," Armstrong muttered.

**Figure 6.14. While it appeared smooth from orbit, the Sea of Tranquility was anything but. During the final phases of the landing, Armstrong struggled to find a safe spot to land. (Courtesy of NASA.)**

"200 feet [61 m]," Aldrin continued. "5½ down..." referring to their rate of descent.

Armstrong pressed forward. There had to be a flat spot out there somewhere. They were so close.

He later recalled, "There's a lot of concern about coming close to running out of fuel, and I was very cognizant of that. But I did know that if I could have my speed stabilized and attitude stabilized, I could fall from a fairly good height, perhaps maybe forty feet or more in the low lunar gravity, the gear would absorb that much fall. So I was perhaps probably less concerned about it than a lot of people watching down here on Earth."[55] Less concerned, perhaps—but focused on his piloting like never before.

The fuel was now down to the point at which the measuring device could no longer sense how much was left—the "Low Level" light came on in the LM and down at Mission Control. Aldrin said, "One hundred feet [30 m] ..." then, "Five percent," which was the estimated remaining fuel.

In another minute and a half, they would have to be setting down or punch the abort button, shedding the descent stage and heading back up to orbit, according to mission rules.

Armstrong still had to stop his horizontal movement and descend— any sideways motion could tear off a landing leg or cause the LM to topple to the side, dooming the crew. But he was still hovering and flying forward, looking for a smooth place to set down the lander.

They had long ago given up trying to identify landmarks. Nothing matched the maps at this close range.

The radio crackled—"Sixty Seconds." That was the CAPCOM telling them how much time they had left to set down. Armstrong and Aldrin were still seventy-five feet above the moon. The crew was silent, con-centration fixed on the job of setting down. Then Aldrin's voice came through: "Thirty feet, two and a half down. Faint shadow."

"Four forward, four forward, drifting to the right a little," Aldrin continued. Armstrong was silent.

When Mission Control announced, "Thirty seconds ..." Aldrin took his eyes off the computer readout for a moment to check the location of the abort button. One of them might be using it within seconds.

Armstrong was trying to cancel the LM's sideways drift to bring it straight in—he had briefly sighted a landing spot. He couldn't actually see it now, as the exhaust from the descent engine was kicking up lunar dust that had sat, undisturbed, for over a billion years, obscuring the view below. But he could see their shadow against the dust billows, and that was enough. "At that altitude, running out of fuel wasn't a consideration," he later said, "because we would have let it just quit on us, probably, and let it fall on in."[56]

A few seconds later, Aldrin said, "Contact light." A five-and-a-half-foot metal rod that extended below three of the LM's landing legs had touched the lunar surface, which triggered a light on the instrument panel. "Shutdown." Armstrong said. Aldrin added, "Okay, engine stop." It was 102 hours, 45 minutes, and 43 seconds since launch.

At Mission Control, the CAPCOM said, "We copy you down, *Eagle*."

Then, from Armstrong: "Houston, Tranquility Base here. The *Eagle* has landed."

Upon landing, less than an estimated minute's worth of fuel settled into the tanks. It had been a close thing. Armstrong later commented, "I thought that the lunar descent on a [one-to-] ten scale was probably a thirteen."[57]

Armstrong and Aldrin quickly went through an emergency liftoff checklist—which they might have used, had the icy fuel-line plug that was silently building up pressure in the landing engine not melted quickly (see chapter 4). Then they had some personal time. Both astronauts marveled at the view outside—*Eagle* had set down on a broad, flat plain, blasted with craters of varying sizes and with rocks and boulders scattered about. In the distance were some low ridges, about twenty to thirty feet high. Armstrong allowed himself a sigh of relief that they were down and safe.

**Figure 6.15. The first image taken by humans on another world—Armstrong took a series of images, assembled here, out the window of the Lunar Module shortly after landing. (Courtesy of NASA.)**

Neither of them had any idea of where they actually were, other than that they were at the end of a long landing zone on a map created from orbit in previous flights. Armstrong said wryly, "The guys that said that we wouldn't be able to tell precisely where we were are the winners today."[58]

Aldrin then said:

> It looks like a collection of just about every variety of shape, angu-larity, granularity, about every variety of rock you could find. The color ... varies pretty much depending on how you're looking rela-tive to the zero-phase point. There doesn't appear to be too much of a general color at all. However, it looks as though some of the rocks and boulders, of which there are quite a few in the near area ... it looks as though they're going to have some interesting colors to them.[59]

The geologists in the back room in Houston listened with close atten-tion. Aldrin's detailed, almost clinical observations would be invalu-able to them.

There was then some technical housekeeping to do, followed by a "rest period," a scheduled nap that seemed absurd now that they were here, but which had been inserted into the flight plan at the insistence of the doctors. The astronauts would happily take a rest, but neither was interested in trying to sleep now that they were on the moon. They wanted to explore.

Aldrin wanted to take a moment to observe his faith, but on a pre-vious flight, that of Apollo 8, when the crew had read from the book of Genesis on Christmas Eve, NASA had received complaints from the public. So Aldrin said, simply, "This is the LM pilot. I'd like to take this opportunity to ask every person listening in, whoever and wherever they may be, to pause for a moment and contemplate the events of the past few hours and to give thanks in his or her own way. Over."[60] In his personal kit he carried a tiny chalice, a few ounces of wine, and a com-munion wafer. He mouthed a few words off microphone and ate the wafer as Armstrong looked out the window.

Then the two of them rested and chatted a bit while preparing to go outside earlier than planned with the agreement of Mission Control.

After a couple of hours, they started the long process of donning their EVA suits and life-support backpacks, checking each item as they went, then cross-checking each other's. The backpacks, called Portable Life Support Systems, or PLSS packs, were a miracle of 1960s technology. The PLSS packs could support a human on the lunar surface for hours using oxygen, carbon dioxide scrubbers, and a water supply both for drinking and keeping the astronauts at a comfortable temperature—the water ran through tiny tubes sewn into their undergarments. Outside, their suits would be exposed to an approximately 500-degree Fahrenheit (260°C) differential between their sunlit side and the shadow side, and cooling was critical.

These tasks completed, the astronauts depressurized the cabin and went to open the LM's front hatch. This proved to be a challenge—despite having vented the atmosphere from the LM, there was still enough residual air inside that the inward-opening door refused to budge. They struggled with it for a few minutes.

"We tried to pull . . . the door open and it wouldn't come open. And we thought, well, I wonder if we're gonna get out or not? And it took an abnormal time for it to finally get to a point where we felt we could pull on a fairly flimsy door and, you don't [want to] rupture that door and leave yourself in a vacuum for the entire rest of the mission back up through rendezvous and everything else," Aldrin later said. "You want to be a little careful about not bending that door. But it did take us an unexpectedly long time to get that, that last tenth of a psi or hundredth of a psi, all of that pressure being pushed against the door."[61]

Eventually, Aldrin grabbed the corner of the hatch and flexed it ever so slightly until the remaining air vented outside with a hiss. The hatch swung inward, and the moon—a bright gray and tan, beckoned to them.

Armstrong got down on all fours, maneuvering carefully in the bulky EVA suit, and started out feet first. At some point, he apparently snapped off a protruding plastic switch on one of the LM's control panels—it was the arming switch for the ascent engine. This went unnoticed at the time but would come to their attention later.

Care was the watchword, despite the broken switch. "The LM structure was so thin one of us could have taken a pencil and jammed it through the side of the ship," Aldrin recalled. "We felt like two fullbacks trying to change positions inside a Cub Scout pup tent," he added.[62]

Aldrin talked Armstrong through the hatch. "All right. Move . . . to your . . . roll to the left. Okay. Now you're clear. You're lined up on the platform. Put your left foot to the right a little bit. Okay. That's good. Roll left. Good."

Armstrong crawled into a small platform outside the hatch, called the "front porch," and pulled on a handle to release the TV camera mounted to a side panel on the LM. A ghostly image replaced the static on Mission Control's big TV screen—it was upside down at first, but the throw of a switch at a tracking station corrected this. He then descended the ladder.

At the bottom rung he made a few experimental jumps back up to assure himself that they would be able to get back inside the LM once they had completed their EVA—there were a few feet between that last rung and the surface.

He was now standing on the footpad. This was the moment.

Armstrong steeled himself, and said, "I'm going to step off the LM now."

He stepped onto the moon. "That's one small step for man, one giant leap for mankind," was heard at Mission Control and across the world, with an estimated audience of about 500 million people watching raptly, many with tears in their eyes. It was the culmination of a decade of excruciating, round the clock effort, the combined work of nearly 400,000 people on various government-supported payrolls across America. Internationally, the United States was at a peak of popularity not seen since the end of World War Two. Two humans—Americans—were on the moon.

Armstrong later noted that he had intended to say, and thought he had said, "That's one small step for *a* man," otherwise, it would not have made sense to his way of thinking. In the intervening decades, various people, including academics studying the moment, have attempted to hear any dropouts or glitches in the transmission that might have indicated that the article "a" was lost, but in the end, it is impossible to say. But the words we heard on Earth were poignant enough.

Armstrong stood just beyond the footpad, surveying the lunar surface with his gold-plated sun visor pulled up to allow him to see the details. "The surface is fine and powdery," he said. "I can kick it up loosely with my toe. It does adhere in fine layers, like powdered charcoal, to the sole and sides of my boots. I only go in a small fraction of an inch, maybe an eighth of an inch, but I can see the footprints of my boots and the treads in the fine, sandy particles."[63]

Aldrin then passed down a camera to him, using a cord that was rigged like a conveyor belt. Armstrong began snapping photos of his surroundings.

This caused some confusion at Mission Control, where people were looking at the timeline checklist for the EVA. The first goal upon reaching the surface was *supposed* to be the gathering a contingency sample, a bit of lunar rock soil near where Armstrong stood. If for some reason they had to depart in a hurry, this would be the only tangible prize from the landing. But Armstrong, the commander of the mission, felt that since this side of the LM was in deep shadow, he wanted to shoot the photos in that light. Not knowing this, Houston reminded him to get the sample as soon as possible. "Roger," Armstrong replied, "I'm going to get to that just as soon as I finish these picture series."[64]

He finished taking the pictures while Aldrin prepared to come down the ladder. With the same caution exhibited by Armstrong (and not breaking any switches in the LM's cabin), Aldrin descended. When he stepped off the footpad and turned to see the landscape before him, he was stunned.

"Beautiful view!" he said.[65]

"Isn't that something?" Armstrong replied. "Magnificent sight out here."

Aldrin paused, then uttered what is perhaps the most hauntingly poetic remark ever made about the bleak splendor of the lunar surface made by any astronaut who visited there. It came out as a simple statement: "Magnificent desolation," he said, almost dreamily.

Armstrong was almost chatty about what he was seeing—"This has a stark beauty all its own. It's like the high desert of the United States," he said, adding to Aldrin's comment. In a later interview, he added, "I was surprised by a number of things ... by the apparent closeness of the horizon. I was surprised by the trajectory of dust that you kicked up with your boot, and I was surprised that even though logic would have told me that there shouldn't be any, there was no dust when you kicked. You never had a cloud of dust there. That's a product of having an atmosphere, and when you don't have an atmosphere, you don't have any clouds of dust."[66]

Aldrin too was marveling at the characteristics of lunar soil, scuffing at it with the toe of his boot, an activity that led him to take one of the most famous pictures in the history of photography:

> The fascination that I immediately had with the fine dust just prompted me to want to record that and to see what it's behavior was. And so I kicked my foot down and, and the response of the dust was different ... it sent little clouds up, not billowing at all, because there's no air, and it would go out and it would all seem to land in the semicircle based on the angle of departure and, and the angle of coming down. And I felt, gee, we need to record this fineness. So had a little insight, well, I want a before and an after, so I took a picture of flat surface. Then I stuck my foot down and pulled it away, took a picture of that. And that looked so lonesome to me, but how can I have a foot and the boot print in there, so I put my foot down another time and then moved my foot away slightly so you could see the boot.[67]

Armstrong and Aldrin had just over two hours to complete this first exploration of the moon—far too little time to do everything they would have liked. Later expeditions would have two, then three moonwalks scheduled, and also carry an electric rover to transport the astronauts and to extend their explorations. But this first mission was kept deliberately simple, and time was at a premium, so they got to work.

First, they held a brief ceremony. The pair moved back to the ladder of the LM to unveil a plaque on the front leg of the lander. Armstrong read the inscription:

> For those who haven't read the plaque, we'll read the plaque that's on the front landing gear of this LM. First there's two hemispheres, one showing each of the two hemispheres of the Earth. Underneath it says, "Here men from the planet Earth first set foot upon the Moon, July 1969 A.D. We came in peace for all mankind." It has the crew members' signatures and the signature of the President of the United States.[68]

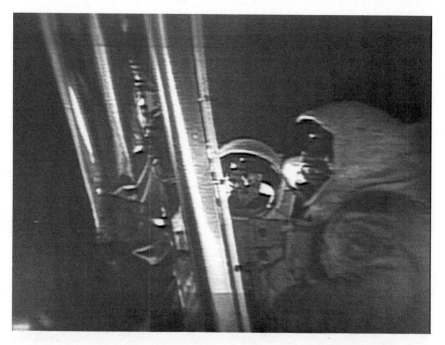

**Figure 6.16. Armstrong and Aldrin read the dedication plaque on the Lunar module. It ended with the words, "We came in peace for all mankind." (Courtesy of NASA.)**

They then set up the TV camera on a stand to transmit their activities, and began working through the checklist that had been so laboriously assembled to best utilize their time on the moon.

Aldrin set up an experiment called the Solar Wind Collector that hung a sheet of thin metal foil on a frame to collect ions streaming from the sun. Without Earth's atmosphere, and well beyond our planet's protective magnetosphere, this would be a physical recording of the effects of exposure of metal to raw solar energy striking the moon.

Minutes later, the astronauts planted the US flag on the lunar surface. This was made of nylon and mounted on a crossbar that hinged from a metal flagpole. The crossbar was needed to suspend the flag since there is no air on the moon. The two of them pounded it as far into the soil as they could, but the flagpole was still a bit wobbly—it would later fall to the ground when the lunar module departed, but it stood up during the moonwalk, which was all that mattered.

**Figure 6.17. Television image of Armstrong and Aldrin erecting the flag on the moon. (Courtesy of NASA.)**

After a few more activities, they received a phone call through a link from Mission Control—President Nixon wanted to congratulate them while they were on the moon.

"Hello, Neil and Buzz," Nixon said:

I'm talking to you by telephone from the Oval Room at the White House, and this certainly has to be the most historic telephone call ever made. I just can't tell you how proud we all are of what you have done. For every American, this has to be the proudest day of our lives. And for people all over the world, I am sure they, too, join with Americans in recognizing what an immense feat this is. Because of what you have done, the heavens have become a part of man's world. And as you talk to us from the Sea of Tranquility, it inspires us to redouble our efforts to bring peace and tranquility to Earth. For one priceless moment in the whole history of man, all the people on this Earth are truly one; one in their pride in what you have done, and one in our prayers that you will return safely to Earth.[69]

There was a long pause, and then Armstrong responded, "Thank you, Mr. President. It's a great honor and privilege for us to be here representing not only the United States but men of peace of all nations, and with interests and the curiosity and with the vision for the future. It's an honor for us to be able to participate here today."[70]

And then, with millions of dollars of moon-time spent on the public relations needs of the executive office, the pair went back to work.

Armstrong trotted away from the LM to collect more rocks and soil. This was called the bulk sample, and was to be gathered as far from the LM as possible to avoid contamination from the rocket plume and disturbance from the exhaust. Armstrong recalls making many short treks to find representative samples. As he said in a mission debriefing, "I probably made twenty trips back and forth from sunlight to shade. I took a lot longer, but by doing it that way, I was able to pick up both a hard rock and ground mass [soil] in almost every scoopful. ... This was at the cost of probably double the amount of time that we normally would take for the bulk sample."[71]

A professor of geology from the California Institute of Technology, Lee Silver, had been recruited to train the astronauts about geological sample collecting, and was impressed. "What Neil did in the shortest period of time that anybody [had] was so brilliant from this point of view of providing the materials to the scientists, that nobody can claim to have exceeded it in production per minute. He was really out-standing." Silver was impressed that Armstrong was actually breaking mission rules to collect the best samples possible. NASA had given him "a very strict protocol," Silver recalled, "which said, 'You will never leave the field of the [TV] camera.' Neil Armstrong recognized that just beyond the field of the camera was a rim of craters covered with rocks and dust, which had been excavated from a little deeper than everywhere else, and he had a very special box for bringing back good samples with a special seal on it, and for about seven or eight minutes, you couldn't see Neil."[72]

In the meantime, Aldrin prepared to set up an experimental package called the Early Apollo Scientific Experiment Package, or EASEP. It consisted of a seismometer to measure "moonquakes" (if any), a lunar dust detector, and a radio for communicating with Earth.

By the time he was done, the CAPCOM told them that they had been out on the surface for two hours and twelve minutes, just a few minutes shy of the planned duration of the EVA. The astronauts grunted their reception of the message. Houston then added an extra fifteen minutes beyond the scheduled time in an effort to accomplish as much as possible. "Okay. That sounds fine," said Armstrong.

Aldrin recalled, "Both of us were concerned with the next item on the agenda to be accomplished. . . . We understood the significance of what we were doing. I felt like we were not alone—we had people lis-tening and looking at everything we were doing, and I had the impres-sion of being on center stage during the entire operation. It was a full schedule of activities."[73]

Armstrong still had more rock and soil gathering on his checklist, called the documented sample. This involved photographing and doc-umenting each item he picked up to provide context for the geologists once it was returned to Earth. At this point the astronauts were no longer walking or slowly hopping across the moon, they were running in long, low-gravity lopes to accomplish their tasks. Lunar soil sprayed from their boots as they dashed from one activity to another to finish on time.

As he was completing his geological assignments, Armstrong wished he knew more about the discipline: "Some [astronauts] had the opportunity and the interest to become very good geologists. I never put myself in that category. I enjoyed geology, and it was certainly appropriate to understanding what we were seeing on the surface of the Moon, but our time was quite limited there. We had a lot of things to do. Had I been a better geologist, I might have seen some things that were important, that I missed. If that's true, I regret it."[74] He added that he thought later crews did a better job of lunar sampling, but with only a couple of hours outside the LM, Armstrong did an admirable job.

Meanwhile, Aldrin started to collect some core samples. The geologists were quite keen to get a sample of material from beneath the surface; the deeper the better. This involved pounding a hollow tube into the soil, which would pull up some of the deeper history of the moon—but this was more difficult than anyone had anticipated; Aldrin was struggling to drive the tube as deep as he could with the use of his geology hammer. "I hope you're watching how hard I have to hit this into the ground, to the tune of about 5 inches, Houston," he said, continuing to pound the sample tube.[75] To gain a depth of less than a half foot, he was forced to raise his hammer above his helmet to get enough force to drive deeper.

"I found that wasn't doing much at all in the way of making it penetrate further," Aldrin said later. "I started beating on it harder and harder, and I managed to get it into the ground maybe 2 inches more. I found that, when I would hit it as hard as I could and let my hand that was steadying the tube release it, the tube appeared as though it were going to fall over. It didn't stay where it had been pounded in. This made it harder, because you couldn't back off and really let it have it. . . . I was hammering it in about as hard as I felt I could safely do it."[76] Even with the addition of an electric-drill drive to the toolset in later missions, obtaining core samples continued to challenge all the Apollo crews on the moon.

Even with the short extension granted by the Flight Director, Armstrong and Aldrin felt that they were not accomplishing as much as they would have liked—a message came up from Mission Control reminded them once again that time was short. "Neil, this is Houston," said the CAPCOM. "We'd like you all to get two core tubes and the solar-wind experiment; two core tubes and the solar wind. Over." It was an instruction to start gathering things that would be carried back to the LM and to get ready to go back inside. Armstrong responded with a curt, "Roger."

**Figure 6.18. Aldrin struggles to pound the core sample tube deep enough to collect soil from well beneath the lunar surface. (Courtesy of NASA.)**

Just to make sure the busy astronauts got the message, the CAPCOM added, "Buzz, this is Houston. You have approximately three minutes until you must commence your EVA termination activities. Over."

At that moment, Armstrong was struck by how little they had accomplished given what kinds of exploration was possible on the moon. "There was just far too little time to do the variety of things that we would have liked to have done. [There were] rocks in a boulder field out Buzz's window that were 3 and 4 feet in size—very likely pieces of lunar bedrock, and it would have been very interesting to go over and get some samples of those. We have the problem of a five-year-old boy in a candy store. There are just too many interesting things to do."[77]

Reluctantly, the astronauts began wrapping-up their exploration. They gathered the boxes of rocks and soil that had been collected, dusted off their suits as best they could, and then Aldrin headed up the ladder.

Meanwhile, Mike Collins, who had been orbiting overhead and listening to what parts of the moonwalk he could through a relay from Mission Control, was performing a series of experiments and photographing the surface below. He had spent many hours attempting to locate the LM with the onboard navigational telescope without success, and had then settled into a busy routine of planned experimental activities. Soon he would be preparing to rendezvous with the LM when his crewmates returned from the moon. This preparation was more involved than one might think—many contingencies had been planned, including the possibility that he would have to lower his orbit significantly if the LM's ascent stage was for some reason unable to reach the proper altitude for rendezvous.

Once Aldrin was inside the LM, Armstrong passed the lunar samples up to the Aldrin using the conveyor belt. It was challenging, even in the light lunar gravity—the conveyor system was balky; it had obviously never been tested in the lower gravity of the lunar environment. moon dust was sprayed everywhere, including onto Armstrong's suit and the LM's interior, as Aldrin dragged the boxes of rocks and soil into the lander.

With the loading of the samples complete, Armstrong headed up the ladder. He crawled inside without issue, and, just over two and a half hours after they had opened the hatch, it was closed again. "Okay, the hatch is closed and latched, and verified secured," Armstrong said.

As the LM repressurized with oxygen, the two astronauts noted with a chuckle that their pressure suits had not caught on fire. Long before their mission began, a physicist from Cornell University, Thomas Gold, had suggested that any lunar dust that was exposed to oxygen might burst into flame due to chemical reactions. He had also suggested that the moon might be so deeply covered in lunar dust that the LM could sink out of sight upon landing. Gold's opinions carried some weight at the agency—he had been a part of the planning process since before the first days of Apollo—but some of his ideas strained credulity with the other geologists and mission planners. In the end, neither hypothesis was correct. The astronauts did notice, however, that lunar dust—which now covered the interior of the LM in thin layers—had a peculiar smell, one that Aldrin likened to spent gunpowder.

For the next thirteen hours, the pair would eat, rest, reflect on their adventure outside, and prepare to depart. Aldrin took a photo from the LM window of the work area below. "Houston. Tranquility Base. We're in the process of using up what film we have." Aldrin snapped off the remaining shots.

Once they were sufficiently rested, they put their helmets and gloves back on and connected their suits to the LM's life support system, depressurized the LM, reopened the hatch, and tossed out anything not needed to complete their mission. Out went the PLSS backpacks, the cameras, even empty food containers; anything not bolted down and essential to their ascent (other than the fifty pounds of collected rocks and soil, of course) was tossed to lighten the load. The base of the LM soon looked like a high-tech trash pile.

With the hatch closed, the pair repressurized the cabin once again and got ready to sleep in preparation for the demanding ascent to orbit. Armstrong drifted into slumber quickly, but Aldrin, who was closer to a noisy pump in the LM, slept only fitfully.

Eight hours later, the pair were busy going through their checklists as they prepared to leave the moon. One task that was not on any checklist was the broken breaker switch, the one that was needed to arm the ascent engine and had snapped off when bumped by Armstrong's backpack as he was leaving the LM. Aldrin had noticed it after he returned from the EVA, and reported it to Mission Control. This prompted a group of worried technicians to hurriedly create a workaround that would allow the crew to arm the rocket below them without the use of that switch, but the ever-practical Aldrin circumvented this by simply jamming the end of a felt-tip pen into the broken breaker to close it, preparing the ascent engine to fire. The problem was solved for less than a dollar.

At 124 hours, 21 minutes into the mission, Aldrin counted down: "Nine, eight, seven, six, five, abort stage, engine arm, ascent, proceed."[78] Armstrong keyed the computer, and the little explosives below them opened the valves on the helium tanks, allowing the compressed gas to force the explosive chemicals into the ascent engine to do their job. The appropriate explosive reaction occurred. The wire guillotine cut the connecting harness between the ascent stage and the descent stage, and the explosive bolts were severed. Both men breathed a sigh of relief as the ascent stage streaked into the skies.

As they ascended, Aldrin took a moment to look out the window. "We're off. Look at that stuff go all over the place," he said as bits of Mylar fluttered about and the exhaust plume scattered the detritus surrounding the LM's descent stage—and blew the flag over. "Look at that shadow. Beautiful."

"The *Eagle* has wings," Armstrong said once again with joy.

They were soon back in the desired sixty-mile high orbit, and just over three hours later were approaching *Columbia*. The rendezvous would occur behind the moon, and Mission Control would have to wait until the spacecraft cleared the limb of the moon to know if all had gone well. The Mission Control announcer said, "This is Apollo Control; 127 hours, 50 minutes ground elapsed time. Less than a minute now away from acquisition of the spacecraft *Columbia*. Hopefully flying within a few feet of it will be *Eagle*. Docking should take place about 10 minutes from now, according to the flight plan. However, this is a crew option matter. We're standing by for word that data is coming in from the two spacecraft."[79]

Docking complete, Collins assured himself that the seal was airtight and the hatches of both craft were opened. The boxes of lunar samples, film magazines, and other paraphernalia were passed to Collins, who stowed them in storage areas within *Columbia*. Two hours later they closed *Columbia*'s hatch for the last time and jettisoned the LM's ascent stage, which drifted slowly away as they watched. After reporting technical data to Houston, Collins said, "There she goes. It was a good one."[80]

With his crewmates safely back inside *Columbia*, Collins's fears about their successful return were erased.

Five hours later, Collins fired the CM's main engine to depart lunar orbit and begin the long journey home. Relaxing a bit now that the engine had fired—the last major milestone to send them home—Collins said, "Yes. I see a horizon. It looks like we are going forward."[81] The others laughed. "It is most important that we be going forward," Collins continued. Aldrin added, "Let's see ... the motor points this way and the gases escape that way, thereby imparting a thrust that-a-way." They were indeed headed in the proper direction—to home.

As they neared Earth, Mission Control passed along questions from the geologists about the lunar samples. "With respect to the documented sample container—on television it appeared to us as though the samples for that container were in fact being given—being selected in accordance with some thought or consideration being given to the rocks themselves. And we were wondering if you could give any further details from memory about any of these samples, and the context of the material or the surface from which they were taken. Over."

Armstrong replied,

Yes. You remember I initially started on the ... side of the LM that the TV camera was on, and I took a number of samples of rocks on the surface, and several that were just subsurface—15 to 20 feet [5 to 6 m] north of the LM. And then I recalled that that area had been probably swept pretty well by the exhaust of the descent engine, so I crossed over to the southern side of the LM and took a number of samples from the area around the elongate double crater that we commented on, and several beyond that, and tried to take as many different types of rock as I could see by eye, as I could in the short time we had available. There were a number of other samples that I had seen earlier in our stroll around the LM that I had hoped to get back and pick up and put in the documented sample, but I didn't get those and I'll be able to comment on those in detail when we get in the debriefing session.[82]

Mission Control acknowledged, and seemed content to await the samples back at the Lunar Receiving Laboratory in Houston. Reentry was just hours away.

Just prior to reentry, Collins released the Service Module (the power and propulsion unit) from the Command Module, leaving just the capsule itself to return to Earth. Reentry was coming up in just a few minutes.

*Columbia* plunged into the Earth's atmosphere on July 24th. The heat shield charred as it heated to upwards of 5,000 degrees Fahrenheit. The reentry speeds encountered when returning from the moon were much higher than those occurring from an Earth-orbital reentry, with resulting higher temperatures. This was just the third spacecraft to come home at lunar speeds, and the tension at Mission Control was palpable—returning spacecraft are unable to communicate with the ground while engulfed in the hot plasma cloud generated by their plunge through the atmosphere.

Sailors aboard the recovery ships stationed in the Pacific Ocean scanned the skies for the three parachutes that would slow *Columbia*'s decent, while all eyes at Mission Control were riveted to the video coming in from the big screen, which showed televised images coming in from the navy. The minutes ticked by slowly . . .

A few minutes later, the parachutes were sighted, and then, after a couple of technical transmissions to the nearby aircraft carrier *Hornet*, Armstrong said, "Hello, *Hornet*. This is Apollo 11 reading you loud and clear."

*Columbia* descended into the Pacific, splashing down 195 hours after launch, thirteen miles from the *Hornet*. Navy divers were dropped by helicopter within minutes, affixing floatation devices to the capsule.

Once *Columbia* was stable and with the recovery helicopter hovering overhead, the hatch was opened and three biological containment suits were tossed in, which the astronauts quickly donned. There had been concerns in the biomedical community that the crew might be carrying dangerous organisms from the moon, and to allay this concern, the astronauts would be kept in quarantine for about three weeks.

Once they were back at the *Hornet*, the trio walked directly from the helicopter to a modified Airstream travel trailer, which was hermetically sealed, to begin their long wait for freedom. Aboard the ship they were visited by President Nixon, who chatted with them through the glass. Within days they had been returned to Houston, where a larger quarantine facility awaited, as well as their families, and they spent their time between debriefings and the writing of post-flight reports relaxing.

The crew found that they welcomed this time to decompress from the mission before encountering the media, which was nearly hysterical in its appetite for face-to-face interviews with the lunar heroes. While they were all reflective about the experience, Aldrin had an especially wistful insight. Near the end of their isolation, he was watching taped news coverage of the mission, which was strange to see from a third-person, earthly perspective. He'd been on the moon, at the center of it all, yet he felt oddly removed. Aldrin later said that this was the first time he actually felt the emotional impact of that they had done—at one point he turned to Armstrong, who was watching the screen in silence, and said, "Neil, we missed the whole thing."[83] Armstrong just smiled.

After three weeks of confinement, the final leg of which was spent in the Lunar Receiving Laboratory in Houston, where their cache of lunar samples also resided, the astronauts were released to the embraces of family and friends. The quarantine was officially lifted by officials at the Center for Disease Control and Prevention in Atlanta.

On August 13, the astronauts embarked on a their first tour since returning—parades, meet and greets, and formal dinners awaited them in New York, Chicago, and Los Angeles. Confetti filled the Chrysler convertible in which they rode in New York, and in Los Angeles the astronauts and their families dined at the Century Plaza Hotel with President Nixon; his wife, Pat; Ronald Reagan, then governor of California; fifty members of Congress; and representatives of eighty-three countries.

**Figure 6.19. A parade celebrating Apollo 11 makes its way down Broadway in New York City on August 13, 1969. (Courtesy of NASA.)**

The astronauts then departed on a forty-five-day tour of twenty-five countries. At each stop they were greeted with parades, formal dinners, and other activities—the world was as spellbound as the United States by the three unlikely heroes. It was at once thrilling and exhausting—while they had known there would be some fanfare upon their return, none were prepared for the outpouring of raw jubilation that awaited them. There was one other side effect of the endless public relations activities—they prevented each man from completing his final psychological transition back to Earth. That denouement would have to wait for months.

Finally, after more than two months of activity, each of them was able to return to his home in Houston to ponder the future. Their paths moving forward were surprisingly divergent, with the common theme being that none of them would fly into space again.

Neil Armstrong had decided long before that Apollo 11 would be his last mission for NASA. He'd spent years in training and weeks in space, and that was enough. He further felt that it was time to step aside and allow other astronauts their turn—he was that generous of spirit. Armstrong had also had enough of the limelight.

Within two years, he accepted a post as a professor of aeronautical engineering at the University of Cincinnati. He took readily to the quiet anonymity of academic life and avoided the press. He said that he simply wanted to be treated as "Mr. Average Guy," wrote Al Kuettner, a representative of the university.[84] UC professor Ron Huston said, "Neil viewed himself as just an ordinary person. . . . He fully understood that the moon landing was the result of [the] long, hard work of many people. Neil did not want to leave the impression that he did it all on his own."[85]

Of his professorial career, Armstrong once said, "I take substantial pride in the accomplishments of my profession. . . . Science is about what is; engineering is about what can be."[86] He died of complications from cardiovascular surgery in 2012.

Of his years with the space program, and in particular the short decade of Apollo, Armstrong said: "I think it's predominantly the responsibility of the human character. We don't have a very long attention span . . . we have a difficult time looking very far into the future. We're very 'now' oriented." He added, "I think we'll always be in space, but it will take us longer to do the new things than the advocates would like, and in some cases it will take external factors or forces which we can't control and can't anticipate that will cause things to happen or not happen. Nevertheless, looking back, we were really very privileged to live in that thin slice of history where we changed how man looks at himself and what he might become and where he might go."[87]

Mike Collins had a more outgoing personality than Armstrong, but he did not seek publicity either. After leaving NASA he spent a year at the US Department of State, but then took a post as the director of the Smithsonian's National Air and Space Museum (NASM). He oversaw a massive expansion of the museum that opened on the American bicentennial in 1976. In 1980 he left the Smithsonian to enter business. Now retired, Collins has the life of a true raconteur, painting watercolors and appearing at very select space-themed events.

**Figure 6.20. Neil Armstrong joined John Glenn to celebrate the fiftieth anniversary of Glenn's Mercury flight. (Courtesy of NASA/Bill Ingalls.)**

Of the future of space exploration, Collins said this while addressing a joint session of Congress a few months after his return from the moon:

During the flight of Apollo 11, in the constant sunlight between the earth and the moon, it was necessary for us to control the temperature of our spacecraft by a slow rotation not unlike that of a chicken on a barbecue spit. As we turned, the earth and the moon alternately appeared in our windows. We had our choice. We could look toward the Moon, toward Mars, toward our future in space—toward the new Indies—or we could look back toward the Earth, our home, with its problems spawned over more than a millennium of human occupancy.

We looked both ways. We saw both, and I think that is what our nation must do.

We can ignore neither the wealth of the Indies nor the realities of the immediate needs of our cities, our citizens, or our civics. We cannot launch our planetary probes from a springboard of poverty, discrimination, or unrest. But neither can we wait until each and every terrestrial problem has been solved. . . .

Man has always gone where he has been able to go. It is that simple. He will continue pushing back his frontier, no matter how far it may carry him from his homeland.

Someday in the not-too-distant future, when I listen to an earthling step out onto the surface of Mars or some other planet . . . I hope to hear, "I come from the United States of America."[88]

Were he to make that statement today, he would doubtless be inclusive of both genders in his speech.

Buzz Aldrin's trajectory has been very different, in both its highs and its lows. Those who knew him said he became somewhat withdrawn after the mission, and that he was having difficulty looking into the future. Aldrin had strong convictions about America's future in space, views he holds to this day—he feels that we have not lived up to our full potential. This is a subject that he writes and speaks on frequently, and he is one of the few space race–era astronauts who has continuously campaigned for a more robust program since the end of Apollo, emphasizing human Mars exploration and international cooperation.

Aldrin later became a media figure, appearing on everything from *The Carol Burnett Show* in 1972 to *The Big Bang Theory* in 2012, with over 150 media appearances all told. Says Roger Launius, a former space-history curator at NASM, "He doesn't stop. He could if he wanted." But he hasn't. He sits on numerous boards and is a member of many professional organizations, working to advance the cause of spaceflight.

**Figure 6.21. Mike Collins chats with Buzz Aldrin at Neil Armstrong's memorial service in August 2012. (Courtesy of NASA.)**

Armstrong, too, had his opinions about the path ahead, and what it will take to accomplish. Reflecting on the national commitment to Apollo, and what it will take to achieve such programs in the future, he said:

I can only attribute [our success] to the fact that every guy in the project, every guy at the bench building something, every assembler, every inspector, every guy that's setting up the tests, cranking the torque wrench, and so on, is saying, man or woman, "If anything goes wrong here, it's not going to be my fault, because my part is going to be better than I have to make it." And when you have hundreds of thousands of people all doing their job a little better than they have to, you get an improvement in performance. And that's the only reason we could have pulled this whole thing off.[89]

Looking ahead to the future, he said,

Mystery ... is a very necessary ingredient in our lives. Mystery creates wonder, and wonder is the basis for man's desire to understand. Who knows what mysteries will be solved in our lifetime, and what new riddles will become the challenge of the new generations?

Science has not mastered prophesy. We predict too much for next year yet far too little for the next ten. Responding to challenge is one of democracy's great strengths. Our successes in space lead us to hope that this strength can be used ... in the solution of many of our planet's problems.[90]

CHAPTER 7

# PETE CONRAD: SALTY SAILOR OF THE SKIES

**Figure 7.1. Charles "Pete" Conrad Jr., about a year after Apollo 12. He was one of NASA's most successful astronauts, and certainly the most entertaining. (Courtesy of NASA.)**

"**H**oly Christmas!"

"Dang it . . ."

"Ho ho ho!"

"By golly . . ."

"Oh for gosh sake!"

"Gosh darn it!"

"Tally-ho!"

In 1973, when Pete Conrad flew as the commander of the first crew to visit America's new Skylab space station there were at least two concerns looming in his mind. His first, and overriding, concern was the task of repairing a damaged solar panel that had gotten hung up during Skylab's tumultuous launch—one panel had disappeared completely, and the other had not unfolded as it had been designed to do. In distant second place, but surely in his mind, was keeping control of his mouth. Conrad was a navy man—sailor, naval aviator, and, later, hotshot test pilot—and was not known for having a dainty vocabulary. To put it mildly, Pete was a potty mouth.

He was also one of NASA's finest astronauts and, in the wild-and-wooly days of the space age, excellence counted.

But NASA also had an image to maintain, one for the organization at large and another for those who were supposed to be squeaky-clean astronauts. There was no small amount of "housekeeping" by the agency's public relations staff when things went off the rails, and salty language was not the worst of their concerns. There were other indiscretions of a more physical nature (many of the Apollo-era astronauts' marriages did not survive long after the program hit its stride). But Conrad was an exception—while his lifestyle stayed well within the boundaries of acceptable behavior for an astronaut, his mouth frequently went off the reservation, with many of his descriptive phrases having far more to do with the unique coupling of body parts and orifices than "Dang it" would imply. Far more.

By the time Skylab flew in 1973, Conrad had flown in the Gemini program (two missions) and landed on the moon on Apollo 12 (the third person to set foot on the moon). He'd behaved well during these missions, keeping his speech relatively family friendly when he was on the air. But Skylab was a whole new animal—almost a month in space for this first crew (there would be three crews in total, each spending more time than its predecessor), and it was a choice mission in the post-moon landing era. But awaiting this first crew was a crippled space station. The possibilities for verbal outbursts from a man famous for a highly creative vocabulary even in test pilot circles were profound. But NASA officials let him know their expectations, and Conrad complied admirably. One can imagine what was actually going through his mind when things went awry in flight—the audiences of 1973 definitely expected G-rated banter from their astronauts.

There had been other verbal excesses of course, perhaps the most famous of which occurred during the flight of Apollo 10. This mission, the last test flight before the landing of Apollo 11, was to do a close-pass flyover of the moon's surface with the Lunar Module, but no landing. Tom Stafford was the commander, with Gene Cernan (who would later land on the moon in Apollo 17) as the LM pilot. When the two men were making their flyover of the moon and separated the LM's ascent stage from its descent stage to return to the Command Module, the LM briefly spun out of control, forcing the astronauts to struggle to regain their orientation. A few choice words were heard on the mission downlink—"Son of a bitch!" among others. This resulted in a reprimand upon their return; never mind that the crew had been in mortal danger, they were expected to keep their potentially final words clean for the audience at home.

So when the time came for Conrad to command the first mission to Skylab, appropriate language was on a few minds.

Skylab was America's first space station. The orbiting laboratory was assembled inside a repurposed upper stage of a Saturn V moon rocket and launched in May 1973. Almost immediately, however, problems made themselves known—something had gone wrong during the launch and, while Skylab made its way into an acceptable orbit, it was generating just a fraction of the power required to operate properly. Within hours, Mission Control came to understand the nature of the problem—for some reason, the two large solar panels on either side were not generating power. It was later realized that one panel had sheared off completely, while the second was stuck to the side of the station, held in place by metal debris that had gotten twisted up as the rocket ascended. It would be up to Conrad and his crew—Joe Kerwin, the Science Pilot, and Paul Weitz, the Command Module Pilot—to attempt repairs to the crippled station. They began preparing for what had become a salvage mission immediately.

NASA could not have chosen a better person to lead this challenging mission. While Conrad had been selected well before the problems with Skylab became apparent—the crew had been training for the mission for well over a year—once the station was in trouble, the selection of Conrad as the leader seemed even more serendipitous. He was endlessly optimistic, boundlessly enthusiastic, highly intelligent, extremely driven to succeed, and a fine astronaut beloved by every crew he flew with. In short, he was a born leader. You could be excused if you did not realize this upon first sighting him, however.

Conrad was nearly bald as an egg, stood five feet, six inches in his bare feet, and was not what you would call Hollywood handsome. He also had a gap between his front teeth that could be seen from across the room, and people saw it frequently due to his boundless sense of humor . . . usually at someone else's expense.

But to judge Pete Conrad by physical appearance, or even his frequent outbursts of sharp-edged humor, would be a disservice . . . and one that you might well pay for by being at the receiving end of his sharp tongue. For he was at once brilliant, driven and passionate, and fiercely loyal to his crewmates, friends, and family.

Conrad had been born Charles Conrad Jr., in Philadelphia, Pennsylvania, on June 2, 1930. Though some who met him later in life might have expected scrappy roots, his family was actually well to do at the time, having made a substantial sum in banking and real estate. But like so many, the Depression destroyed the family fortune, eventually driving them from their opulent home. They relocated to a carriage house paid for by Conrad's uncle on his mother's side. His father left the family soon thereafter.

But Charles Jr., known universally as Pete, was unstoppable. He was a child of action. Household appliances disappeared with regularity, only to reappear after being disassembled, examined, and put back together—and they still worked ... sometimes. He was given to venturing as far and wide as a young man's feet would carry him during the daylight hours, finding places to try brash youthful adventures such as seeing how high he could jump from a tire swing before his ankles failed him upon landing. In short, it was a Norman Rockwell childhood in post-Depression America.

By his mid-teens, young Pete had scored a job at a local airfield in Paoli, about twenty miles outside of Philadelphia. Rather than be paid, Conrad traded his chores for flight time and basic instruction. He also learned about the mechanics of light aircraft, a background that likely added to his passions for both flying and engineering.

Like most young men in formerly influential Philadelphia families of a certain status, he was sent to the Haverford School, the "right place" to be if you wanted to go places in life. Unfortunately, besides an intrinsic streak of rebelliousness, he also had dyslexia, and schoolwork was a challenge—so much so that by eleventh grade he had been expelled.

His mother, now single, and no shrinking violet herself, would not give up on her son. She sent him off to the Darrow School in upstate New York, which utilized a different approach to learning—one that worked for his different educational abilities—and there he flourished. After repeating most of the eleventh grade, he graduated in 1949, having become the captain of the school's football team, despite his diminutive stature.

He did so well at Darrow that he was admitted to Princeton with an ROTC scholarship. While studying aeronautical engineering at Princeton, Conrad's interest in flying continued, and he received an instrument rating in light aircraft. Upon graduation, he was commissioned as an ensign in the US Navy. A year later, he was a fighter pilot and spent several years flying on aircraft carriers as well as becoming a flight instructor.

But Conrad had larger goals, not the least of which was flirting with danger in ever faster machines. He applied to the Naval Pilot Test School at Patuxent River, Maryland, and was accepted, joining with classmates such as future astronauts Wally Schirra and Jim Lovell. He graduated as a test pilot in 1958 and was promoted to captain.

The year was auspicious—this was the same year that NASA was founded. Almost immediately, the new space agency started scouting for its first class of astronauts, who would eventually become widely known as the "Mercury 7." Requirements were: less than forty years old, a bachelor's degree or professional equivalent, 1,500 hours of flying time with qualification for jet aircraft, training as a test pilot, a height of less than five feet, eleven inches, and a weight of less than 180 pounds. Conrad was qualified in all respects, and he was recruited along with more than five hundred others for this exciting new program. The orders came to him in an envelope marked "Top Secret"—what could be more exciting for a thrill-seeking sky jockey than that? Of course he was interested. He and the others selected plunged into the process that would separate the men from the boys, the best from the extremely good—In short, those who had what Tom Wolfe's seminal early space age book *The Right Stuff* referred to in its title.

Sixty-nine finalists were selected, and Conrad was among them. He was shipped off with the other remaining candidates to undergo extensive physical, psychological, and other testing.

Part of this testing was conducted at the Lovelace Clinic in New Mexico, made infamous by Wolfe's book and the 1983 motion picture somewhat liberally based upon it. While rather exaggerated, the movie is a reasonably good general portrayal of the indignities heaped upon the astronaut candidates while at the new aerospace medical clinic. Most of the test pilots undergoing the tests were grudgingly compliant—they wanted to fly in space enough that they would brave whatever the doctors threw at them. A few, such as John Glenn, were gung-ho about the tests; Conrad was decidedly not.

The testing was extreme and impersonal. Nobody in the medical profession really knew what to expect from the rigors of spaceflight, so anything that could be thought up to account for all the possible reactions of the human body was tested—and then some. The astronaut candidates' heads would be strapped to a chair, ice water would be pumped into one ear, and the motion of the eyes measured. What the hell was that all about? In another torture session, a long needle would be jammed into their thumb muscles and charged with electricity to make their hands clench and wiggle, with the results carefully—and silently—recorded. The doctors even sent the men to the lavatory, sample tube in hand, to provide sperm samples. It was all rather demeaning, and the Torquemadas in charge did not seem to care one bit about the "subjects" they were evaluating.

None enjoyed the process, few were comfortable with it, and Conrad, perhaps above all the others, was demonstrative in his rejection of its value. So he got his revenge, and it probably cost him his slot in the first cadre of astronauts.

His first act of rebellion was to decorate his fecal sample. They wanted two of these in short order, but the very stresses of the procedures were messing with Conrad's digestive system. It took him a number of days, despite continued harassment from the lab-smock team, to turn out even one small ball of poo. But rather than just drop it into the sample cup, Conrad tied a small red ribbon around it and delivered it to the staffers with his trademark gap-toothed grin. They were not amused, or as Wolfe put it, "The Lovelace staffers looked at the beribboned bolus, and then they looked at Conrad ... as if he were a bug on the windshield of the pace car of medical progress."[1] It was not a promising start down the road leading to the conquest of the heavens.

Soon thereafter, the crew was transferred to Wright-Patterson Air Force Base in Dayton, Ohio, for psychological testing. If the physical exams had been extreme, many of the cadre—test pilots all—felt that this batch went right off the bullshit meter. Doctors were people that test pilots avoided in general—they could ground a flyer for what were thought to be absurd reasons—but brain doctors, or "shrinks," were the worst of the bunch. They turned out to be Conrad's Waterloo for his first chance at spaceflight.

Sensory deprivation, tests for reactions to the stress of high-altitude air pressure, and the weirdest little puzzles on paper were all part of the regimen. Other tests involved interviews with all kinds of questions about risk-taking and so forth, and one had to answer these very carefully to maintain the appearance of relative normalcy— or whatever passed for "normal" for a test pilot. But the big day came when the astronaut candidates were each handed a blank sheet of paper and asked what they saw—it was a free-association test, one designed to probe the inner characteristics of the examinee's mind. Some said, "A field of snow." Others said, quite rationally, a "blank sheet of paper." Conrad's response? "But it's upside down . . ."[2] Plus ten points for imagination, zero for cooperation—after a moment of turning the paper around to examine it, the doctor knew he'd been had. And so it continued, one rebellious prank after another.

Not long afterward, when Conrad had returned home to Maryland, he got the letter from NASA. Thanks for your interest, but you didn't make the cut. He eventually learned that his evaluation had stated that he was "not suitable for long-duration flight."[3] So be it. Conrad returned to carrier duty and continued flying jets as he had been trained to. But not for long.

The Mercury program was now at full throttle and going swimmingly. NASA finally had its "man in space" to match the Soviet success with Yuri Gagarin, and more would follow in the tiny Mercury capsule. But since President Kennedy's announcement that America was aiming for a manned lunar landing by the end of the decade, it was clear that more than seven astronauts would be needed. The door was opened for astronaut class number two.

By now, Conrad was flying out of the Naval Air Station Miramar, in Southern California, performing tests for a new carrier night-flying system—one of the most dangerous, and nerve-wracking, types of aviation there is. Alan Shepard himself, the hero of the first manned Mercury flight and the man that put the United States back into the space game, approached Conrad and encouraged him to apply to become an astronaut. Into the test gauntlet he went once again, but this time he behaved better. The tests were also far less invasive, now that the doctors knew that a relatively healthy human could survive the rigors of spaceflight. Conrad made the cut this time and was put on active status as an astronaut in 1962, along with eight other newcomers.

With the flight hours accrued during his time off between astronaut applications, Conrad now had over 6,500 hours of flight time under his belt, always a good thing when vying for a coveted spot on a spaceflight mission. It was not long before he was assigned to a prime seat on Gemini 5, just the third crewed flight in the new program. His commander would be Gordon Cooper, one of the original Mercury astronauts, and of all of them, possibly even more imbued with self-confidence than Conrad.

The Gemini flights were a critical proving ground for the Apollo program, and each mission was intended to build experience in space for the next one. The Gemini program had multiple objectives—tests that needed to be achieved in just ten flights before Apollo could fly. These included long-duration spaceflight, rendezvous and docking with another spacecraft, and working in space during EVA. Gemini 5 was intended to test long-duration spaceflight—not just an arbitrary number of days in orbit, but seven full days, enough to shatter the previous Soviet record in orbit. Secondary objectives involved rendezvousing with a small pod that would be ejected from the Gemini spacecraft, and the use of fuel cells instead of batteries for power. Fuel cells are devices that combine hydrogen and oxygen to create power, and they had not yet been tested in manned spaceflight. These would be critical to powering the Apollo spacecraft, as batteries would not last long enough for lunar missions that could take up to two weeks—solar panels were not yet sufficiently developed for the power needs of the Apollo spacecraft.

On August 21, 1965, Cooper and Conrad were sealed inside the Gemini capsule awaiting liftoff, just a few hours away. Cooper, visor up, turned to Conrad, just a couple of feet away. "You ready rookie?" he said—it was Conrad's first spaceflight. Conrad did his best to look terrified, and said, "I'm not sure, Gordo ..." Cooper just stared at Conrad, who played it straight as long as he could, then broke into that goofy grin of his and said, "Gotcha! Light this son of a bitch, and let's go for a ride!"[4]

**Figure 7.2. Conrad during "water egress training," or exiting after splashdown, from a Gemini trainer. (Courtesy of NASA.)**

The rocket left Earth launched atop a Titan II booster at just before nine a.m. EST. The Titan had been developed as a nuclear weapon carrier—an intercontinental ballistic missile, or ICBM. As such, it had not been originally designed to carry humans. Yet here were two of them, strapped inside the tight-fitting Gemini capsule, atop a Titan. Since its selection for the Gemini program, the Titan had been "man-rated," meaning that everything that could be done to make it safe to carry astronauts had been done, but there was still one problem with the missile, an issue that would crop up again with the Saturn V moon rocket. This was called pogo. In brief, the rocket's twin first-stage engines would experience pulsing while they fired, enough to create some high g-forces during launch. Nuclear weapons don't care much about eight g's (the maximum force it was expected to produce in flight during a surge), but soft, squishy human bodies were potentially endangered by this slamming around.

But this was the space race—the great gladiatorial battle between the United States and the Soviet Union for the heavens. Despite the risks, the Gemini crews went up on the Titan, and it was Gordon Cooper and Pete Conrad's turn to fly The Titan did pogo a bit—enough to cause the astronauts' vision to blur and make speaking difficult—but other than that, the effects were manageable. The problem was later resolved with a small engine modification, and Gemini 5 was the last Gemini flight to suffer from pogo. Minutes later they were in orbit, and the first of NASA's long-duration missions—one intended to mimic the flight time to the moon and back—was officially underway.

Their first experiment involved an attempted rendezvous simulation. To accomplish this, later flights would launch an Agena, with which the Gemini crews would rendezvous and dock, but for this first test Gemini 5 instead carried a small device called the Rendezvous Evaluation Pod in its rear section. This was ejected and slowly moved away from the Gemini capsule, and Gordon and Conrad made preparations to chase it down and perform a rendezvous, the first test to mating up two spacecraft in orbit, just as the Apollo spacecraft would have to do in a few short years.

Then the first malfunction of the mission struck, a bit over four hours into the flight. One of the spacecraft's fuel cells, the experimental new power system, suffered a pressure drop—from the expected 850 pounds per square inch to just 65 psi. Commander Gordon decided to shut down the fuel cells, putting the spacecraft on battery power only—power that would last only hours. This necessitated aborting the rendezvous test.

Subsequent evaluation on the ground led the NASA engineers to decide that it was safe to reactivate the fuel cells, so the astronauts did so, testing them with ever heavier power loads. While the rendezvous experiment was no longer on the table, they would at least be able to continue the mission—being on batteries alone would have necessitated ending the flight within the first day.

However, they were still operating at less than peak capacity, and with the reduced power, the capsule became cold, and the crew found it difficult to sleep during their prescribed rest periods. Wakefulness became the norm. The Gemini capsule had just enough room for two men, less than five-feet, eleven-inches, to sit side by side with their shoulders almost touching and their helmets just inches below the hatches. It was like sitting inside a very small sports car—with no windows to roll down for fresh air—for days at a time. The astronauts' legs were jammed into a cubby below their instrument panel, so they could not even stretch. It was about as far from comfortable as one could get—it made flying coach aboard a modern airliner look like the Hilton.

Meanwhile, fellow astronaut Buzz Aldrin, who had a doctorate in orbital mechanics, worked out another method of attempting an orbital rendezvous. With his guidance, on day three the crew attempted to rendezvous with a predetermined point in space—just an empty, theoretical dot. They were successful, and then returned to shivering the hours away as their tiny capsule continued to orbit the Earth.

On day five, a malfunction of the maneuvering system was detected. Many possible causes were suggested by the engineers at Mission Control and elsewhere, but it was determined that the problem could get worse, and maneuvering was minimized. A number of experiments involving maneuvering and attitude change of the spacecraft were canceled as a result, and Gordon and Conrad began to succumb to the bane of the Gemini program—boredom. As Conrad would later say, it was "just about the hardest thing I've ever done."[5]

Nonetheless, they had other chores to attend to. Photography of the stars and the Earth were of primary interest, and they took to this task with aplomb. Various medical experiments were also undertaken, one of which involved Conrad placing inflatable cuffs around his legs—no easy feat in the tiny cockpit. But it was better than just sitting there.

At one point mid-mission, a bag of freeze-dried shrimp got away from them, and the tiny pink crustaceans began to float all over the cabin. Cooper and Conrad did everything they could to scoop up the pieces of stinky detritus, but through the end of the mission miniscule bits of shrimp and pink dust drifted through the capsule, and even into their lungs.

**Figure 7.3. Conrad waits out the long-duration Gemini 5 mission. With power below expected levels, he and Gordon Cooper spent much of their time in orbit shivering and trying to sleep. (Courtesy of NASA.)**

Toward the end of the mission, power and consumables were becoming an ever larger problem, and for the final orbits, with no maneuvering ability, they simply allowed the capsule to drift. The Earth would pass by, then the starry sky, then Earth again—a never-ending, slow kaleidoscope of color, then blackness; or, when on the night side of our planet, dark and darker. It was mind-numbing, so Cooper eventually fashioned shades over the windows. That way, they would at least have the luxury of not enduring sweeps of bright sunlight every few moments.

After just shy of 191 hours cooped up inside the Gemini spacecraft, it was time to come home. Cooper initiated a second set of maneuvering thrusters—the system dedicated to bringing them home alive—and commenced retrofire. He controlled the reentry by hand to test alterations of the capsule's lift and glide by changing its orientation during reentry, the final experiment of the mission. Then they plunged into the Pacific Ocean and awaited the navy helicopters and divers who would release them from their shrimp-stained, stinky home of over a week.

Cooper and Conrad chatted as the capsule bobbed in the swells, trying to decide if they should open the hatches or just wait for the navy divers to get them out. "What do you think, Pete?" he asked. "It's gotta be a hundred thirty in here, Gordo," Conrad responded. "I'm sweating my ass off!" Just then, there was a thump as a diver pounded on the hatch, announcing his arrival. "I can't do this another minute in here," Conrad said. "Open the son of a bitch."[6]

When they arrived on the carrier *Lake Champlain*, the navy crewmen surrounded them, applauding. With cameras rolling, Conrad gave his trademark mischievous grin and tweaked a surprised Cooper on the nose. He was supremely proud of the mission, but later noted that the week in the capsule had been like being in a flying "garbage can."[7]

Nonetheless, the duo had proved that humans could exist safely in space long enough to travel to the moon and back, had shown that rendezvous could work, and, as a bonus, had reclaimed the long-duration spaceflight record from the Soviets. They had managed to overcome equipment breakdowns and malfunctions and shown, once again, the utility of having humans at the controls. Surely Conrad was impressed by these achievements?

"The romance ended fairly quickly," he later recalled of the time in the capsule. "My knees began to bother me . . . I hurt and I didn't want to stay in there," he said. "If they had told me I had to stay up there longer than the eight days, I believe I would have gone bananas."[8] So much for the silk scarves and derring-do of the rakish aviators of a bygone era.

As challenging as the mission had been, Conrad flew Gemini once more, this time as the commander, in Gemini 11, the penultimate mission of the program. The heat was on to accomplish the tasks that would prove everything needed to accomplish the Apollo missions was feasible. Conrad took to the mission with his usual relish, aided by the fact that he would be flying with a dear friend this time, Dick Gordon. Conrad and Gordon had met at the Navy Test Pilot School at Patuxent River years before and had become fast friends. They flew jets together, drank together, and laughed at the same dumb jokes. They were like brothers. Gordon had not made the cut when he first applied to NASA either, but at Conrad's urging had applied yet again for a slot in the Gemini program. When he asked Conrad why he should bother—he had a bit of sour grapes going on after his initial rejection—Conrad said, "Because you miss me."[9]

Gemini 11 would be another test of EVA (spacewalking), to prove that astronauts could do useful work in weightlessness. They had trained in swimming pools to practice, while weighted to neutral buoyancy. So far the EVAs on Gemini had not gone well—astronauts overheated, drifted away from their chores, and generally had trouble accomplishing even simple jobs outside the capsule. By now the Agena was flying, and the Gemini capsules would dock with it, with one of the astronauts leaving the Gemini spacecraft and proceeding hand over hand to the Agena to retrieve experiments—and just to prove that they could get there. This had been generally successful with some trial and error, and NASA wanted to up the stakes with more complex tasks. This would be Gordon's job, while Conrad sat in the capsule, monitoring his friend as he went about his assigned duties.

But getting in and out of the Gemini was harder than one might think. The capsule was cramped and the hatches small. It was a contortionist act just to get in and out of the thing. And then, what would happen if the spacewalking astronaut suffered a problem outside the spacecraft? He might suffer some kind of seizure in space or have trouble with his breathing apparatus or vomit inside his suit, choking on the floating globs. These frightful circumstances, and a dozen other the doctors dreamt up, might prevent the spacewalker from getting back inside the capsule. What should the commander do? The simple answer, and obvious to the other astronauts, is that the other man would *help* him . . . but that was not NASA's plan.

Shortly before the mission, Conrad was summoned to the office of Deke Slayton, the chief astronaut. When Conrad arrived, he saw that Gene Kranz, who would be the flight director for the mission, was there as well. This could only mean something serious.

Slayton began by saying that they wanted to discuss the EVA. He said that it was "ticklish," and could be more challenging than they thought.

"Aw, Dick's kicking it in the pool. He'll do terrific," Conrad responded.

But Slayton and Kranz did not go for Conrad's aw-shucks answer. They were concerned about the survival of the commander, and the completion of the mission. Conrad listened, scarcely able to believe what he was hearing, as the two outlined their procedures should Gordon for some reason not be able to get back into the capsule. "If Dick can't get back in the ship, you have to leave him," Slayton said.

This really rang Conrad's bell. Leaving a man behind, especially his beloved friend Gordon, was unthinkable.

Kranz picked up the discussion. "Under no circumstances—*zero*—are the two of you to be outside the vehicle at the same time. We are clear on this, right, Pete?"[10] There would be a set of snippers included in the flight kit so that Conrad could cut the air hose and tether affixed to his best friend in the world and watch him drift away, rapidly suffocating, as Conrad buttoned up the hatch and made ready to return to Earth alone. Having just killed his closest companion besides his wife. Alone. Conrad agreed, but he was not at all sure about what he would actually do in such a circumstance. As he left the meeting, somewhat stunned, he put the scenario out of his mind and returned to his training.

Gemini 11 launched on September 12, 1966. As was now routine, the Gemini launch had been preceded by an Agena target vehicle, which launched just hours earlier to give them something to rendezvous with. The Agena sailed into orbit, functioning perfectly, and everything was looking good for the upcoming rendezvous and docking.

Conrad took the controls shortly after the capsule reached orbit, and they began to chase down the Agena—the first time this had been attempted in the first orbit. Previous missions had loitered a bit before going after the Agena, but on this flight they were docked within an hour and a half of launch. Conrad then backed out of the Agena's docking adapter and repeated the procedure four more times—rendezvous and docking was becoming routine, just as NASA had hoped.

The crew then prepared for another first: they would fire the Agena's rocket engine and propel them into the highest orbit ever achieved by a manned spacecraft, 853 miles above the Earth at its maximum extent. After reaching this record-setting altitude—a record for orbiting humans that stands to this day—they fired up the Agena again and returned to an altitude of 184 miles to prepare for Gordon's EVA. The pressure for a successful spacewalk was on.

The day after launch, Gordon conducted the first of his two EVAs. It was scheduled to be a two-hour excursion, in which he would depart the Gemini, move to the Agena, attach a hundred-foot-long tether stored on the Agena to the capsule, and perform other tasks. He got the tether attached, but as with previous Gemini EVAs, found his time outside the capsule exhausting. He was huffing and panting, his heart rate was elevated to over 150 beats per minute, and he was sweating like mad with his visor fogging up. Conrad was getting worried, and painfully aware that those dreaded umbilical snippers were sitting just under the console.

"How you doing, Dickie-Dickie?" he said over the radio, trying to sound calm.

"I'll be okay," Dick wheezed.

Conrad pondered a bit longer, then pulled rank, and ordered Gordon back inside after only a half hour.

"I'm calling this one a wrap. Get back in here, and let's go find a beer somewhere . . ." Always the kind voice, but Gordon got the message and struggled back into the capsule with the balky, snakelike tether greatly complicating things—it was all over the place, and not easy to reel in and coil up. It was a close call.

While the EVA was only partially successful, they did manage to separate from the Agena, connected by the hundred-foot tether, and conduct a couple of tests. The first was to see if the Agena would remain stable if pointed "down" toward the Earth while tethered—it did. The second was to spin up the connected spacecraft and twirl them (a bit like a bolo) and create a small amount of artificial gravity. This was partially successful. In both cases, the tether's motions, weaving and causing the two spacecraft to lurch and wobble, were unexpected and troublesome. They soon detached from it and left the Agena.

Gordon's second EVA was merely a "stand-up" spacewalk—just open the hatch and stand in space with his feet on the Gemini seat. This was not tiring; in fact, Gordon found it relaxing. He took a number of photos of both the Earth and targets among the stars, then, after two hours, sat back down inside the spacecraft.

**Figure 7.4. Conrad with Dick Gordon after splashdown of Gemini 11. (Courtesy of NASA.)**

Just shy of three days after they left Earth, Conrad and Gordon reentered the atmosphere and splashed down only about 1.5 miles from their intended position near the US territory of Guam. It was the first reentry performed entirely by the onboard computer. The mission had been successful enough—except for another frustrating EVA—and it laid the ground for the final Gemini flight, during which Buzz Aldrin—after thousands of hours of practice underwater and in the zero-g plane—would perform a flawless EVA. That was on Gemini 12, and with its conclusion, the program had accomplished just about all of the goals set out for it, paving the way for the Apollo program. The first Apollo flight would take place with Apollo 7, an orbital test flight in 1967 that followed a number of unmanned tests of the Apollo hardware. But Conrad would not be on that landmark mission—he had other duties to attend to, as did all the astronauts headed into the Apollo program. They were regularly traveling from one coast to the other, visiting various NASA contractors and overseeing specific systems and subsystems of the spacecraft and associated hardware.

By late 1966, Conrad was assigned to the backup crew of the first Apollo flight scheduled to test the complete lunar landing system— the Command and Service Module (CSM), which included the Apollo capsule and its propulsion and life-support unit, and the Lunar Module. This mission would end up flying as Apollo 9 in 1969, but it was already in the pipeline three years earlier. The situation was a bit more complex than this, however—the Lunar Module's development was considerably behind schedule. The LM was the first machine of its kind ever built, and making it work was incredibly challenging. It was also far too heavy to meet the weight limits of a lunar landing mission, so many months were spent making the LM ever leaner, even after its construction was well underway. Due to this and other factors, the Apollo 8 mission was flown in December 1968, without an LM, orbiting the moon but obviously not landing there. In any event, the prime crew for Apollo 9, which was launched in March 1969, flew their mission, so Conrad's services for that mission were limited to a backup role. He was assigned to the next available slot, as commander of Apollo 12.

Of note, there is conjecture that, had the Apollo 8 flight not been made early (it was originally intended to be the test mission for the LM in Earth orbit), the usual crew rotation timing might have resulted in Conrad being the first man to step on the moon on Apollo 11, but Deke Slayton, chief astronaut, was in charge of crew assignments and held his decision-making cards close to the vest. In any case, if Conrad were disappointed about the change, he was not vocal about it. He was just thrilled to be a part of it all. Besides, his moonwalk would be far longer and more ambitious in scope than that of Neil Armstrong and Buzz Aldrin on Apollo 11.

Conrad and his crew were preparing feverishly for the Apollo 12 flight as Armstrong and Aldrin made the first lunar landing. Like everyone else who had access to a TV or radio, he followed the proceedings with deep interest—it was just simply amazing that it had all worked—then got back to work training for his flight, which would occur just four months later.

His crew was a choice trio. For his Command Module pilot, he was assigned Dick Gordon, his old pal from flight school and Gemini 11, and rookie Alan Bean filled the third slot. Bean had been inducted into NASA in the third group of astronauts along with Gordon, but there the similarities stopped. Where Gordon was a bit of a roustabout like Conrad, Bean tended to be more thoughtful and quiet by nature. He was a fellow naval aviator, however, and a solid test pilot. Regardless of these qualifications, however, prior to his selection to fly on Apollo 12, he had been assigned to what was widely considered a backwater duty, in an office at Houston that was planning follow-on missions to the lunar landing program, the Apollo Applications Program. Of the many plans that came out of this office, which included long-term lunar surface missions and mobile lunar habitats, the only one that was ultimately undertaken was the later Skylab space station. It was a frustrating assignment, and Bean considered the work to be of secondary importance at best. But he gave it his all, and Conrad took note of Bean's efforts. Conrad had first met Bean as his student at the navy test flight school, and while Bean was not quite the born "stick and rudder man" of the same caliber as Conrad, Gordon, and some of the others, he worked hard—very hard. For this reason and others, including the fact that Bean often spoke his mind when not politically wise to do so, Conrad had taken a shine to the rookie astronaut.

When it came time to fill the third seat for the crew of Apollo 12, Slayton asked Conrad who he wanted for the slot, and Conrad said he wanted Bean. But Slayton was not impressed with the idea, and instead gave him another astronaut named C. C. Williams, who was also eminently qualified. Conrad accepted the decision—he liked Williams just fine and would make it work. But in 1967, when Williams was killed in the crash of his aircraft, Bean was still at the top of Conrad's list, and he finally got the crew he had asked for.

Training for Apollo 12 was something new for the simulation supervisors. The SimSups created increasingly complex and nearly impossible scenarios to throw at the astronauts, introducing multiple failures into the simulation until they managed to cause the astronauts in the sim to crash and burn. Many of the astronaut crews grumbled over this and quietly swore a thousand deaths upon the SimSups. But not the crew of Apollo 12. In fact, Conrad's crew were unlike anything the SimSups had seen since NASA started spooling up for the lunar program. This crew took everything in good-natured stride.

If you were listening to the crew of Apollo 11 train, you might hear Aldrin arguing with Armstrong about whether or not to abort a simulated landing that looked bad. Collins might interject a few soothing suggestions, and Armstrong would stoically go about his business. These were professional pilots working through their assigned duties, united in their efforts to prove themselves to the SimSups. Other crews behaved similarly, but more vocally in their antagonism toward their training tormentors.

But not the crew of Apollo 12. These guys were different—they were like brothers.

No matter how complex or even unfair (and often unrealistic) the expectations of the SimSups were, Conrad and Gordon would work through them with unending good cheer and raucous banter, punctuated with Conrad's salty language. Bean would be quieter than his crewmates, but he was also unfailingly optimistic and, as usual, gave it his all. As the mission got closer, the three of them seemed to become even more closely bonded. It was something unique in the Apollo program—this crew truly loved one another.

As the stress increased, and the launch date loomed ever closer, there was little time to do much other than train, eat, and sleep. But as the commander of the next flight to the moon, Conrad was still required to attend some of the many press events held after Apollo 11's historic first landing. An Italian Journalist offhandedly asked him what he was going to say when he set foot on the moon—surely this had been scripted by NASA, she was certain of it. After all, who thought that someone as seemingly colorless as Neil Armstrong could come up with "That's one small step for [a] man, one giant leap for mankind..."[11] Conrad reminded her that this had in fact been Armstrong's own doing, and that his first words were his to choose. "That's up to me, darlin'," he said with his trademark smirk. She did not believe him for a minute. So Conrad said, "I tell you what. Let's you and me decide right here and now what I'm going to say when I set foot on the moon."[12] He extemporized a pithy statement on the spot that apparently left her slack-jawed, and she didn't buy that either. So he bet her $500 (about $3,500 in 2019 dollars) that he would say exactly that, and she accepted the bet.

Then it was back to training at a frenetic pace. The hardware was ready, as was the crew, and with the success of Apollo 11, confidence was high. All too soon, it was time to go.

On November 14, 1969, the three crewmen arose early, ate a low-residue breakfast (nobody liked pooping in space for obvious reasons), suited up, and rode the eight-mile course to the launchpad, where their Saturn V rocket gleamed in the predawn spotlights. There was a light rain, but this was not unusual weather for the cape—weather can change quickly in that part of Florida. Cryogenic fuels were boiling off of the rocket, creating plumes of white vapor alongside the fuselage. The trio tramped from the shuttle bus to the gantry and took the nearly 360-foot ride to the top of the rocket, where their Command Module awaited. A fellow astronaut, who had been going over his checklist, making sure all the switches were in the proper position for launch and all gauges were giving proper readings, disembarked the capsule, allowing the astronauts, already suited up in their bulky pressure suits, to take their places inside. Conrad went into the left-hand seat, reserved for the commander. Gordon took the center couch and Bean the right. They settled in to await liftoff.

**Figure 7.5. The irrepressible crew of Apollo 12, *from left*: Conrad, Gordon and Bean. (Courtesy of NASA.)**

Jack King, the launch control announcer, kept up his dialog for the news feeds: "This is Apollo-Saturn Launch Control, T-minus 18 minutes and 40 seconds and counting; countdown still proceeding at this time, although it is touch and go at this time, we are still not below our minimum margins for launch..." referring to the fickle weather patterns over the Cape.

The weather held, but barely enough to allow the count to proceed. Nevertheless, at T-minus-zero off they went. The launch was just like the simulations, and would be for about forty seconds.

From the bleachers, then president Richard Nixon watched the space program of his former political rival John F. Kennedy take to the skies, a big grin plastered across his face.

Conrad's voice came down to Mission Control: "I got a pitch and a roll program, and this baby is really going!" followed by, "It's a lovely liftoff.... It's not bad at all."[13]

Then, at about forty seconds into the flight, the radio communications began to crackle and break up. In Mission Control in Houston, the numeric readouts on the screens turned to hash.

Conrad felt something lurch in the spacecraft even as the rocket plowed upward.

"What the hell was that?" he asked, a bit less laconically.

"That's a whole bunch of stuff..." said Gordon.

"Roger, we had a whole bunch of buses drop off..." Conrad said, referring to the electrical breakers having suddenly tripped. "There's nothin', what have we got?"

In Mission Control, Gerry Griffin was filling his first stint as an Apollo Flight Director, and his first day on the new job had just turned sour.

Conrad's voice, calm but firm, crackled over the static: "Okay, we just lost the platform gang, I don't know what happened here, we had everything in the world drop out." The platform was the guidance system that allowed them to navigate, and it had just gone wild.

Conrad read off a long list of malfunctions—the control panel warning lights all flashed on at the same time, something they had never seen before, not even in the worst simulations. "I got three fuel cell lights, an AC bus light; a fuel cell disconnect, AC bus overload 1 and 2, main bus A and B out..."

It seemed like the entire Apollo capsule had failed all at once. But at one of the control consoles, an electrical engineer named John Aaron thought he might know what was going on. He'd seen this once before in an electrical equipment test. "Flight, EECOM. Try SCE to AUX," he said. The CAPCOM relayed the suggestion. There was a pause.

"FCE to AUXILIARY, what the hell is that?" Conrad said. "*S-C-E* to AUX..." corrected the CAPCOM. Then Al Bean lit up, "Uh, I think I know what that is..." and he threw the proper switch. The warning lights winked off, and the displays at Mission Control went back to normal, though it was clear something untoward had occurred.

The first stage dropped free, and Conrad said, "I'm not sure what happened... I'm not sure we didn't get hit by lightning."

Indeed they had been. The bad weather, combined with the effects of the giant rocket lunging through the charged clouds, had caused lightning to hit the Saturn V not once, but twice. Most of the breakers in the Command Module tripped, but fortunately, the Saturn V had been designed with a huge secondary computer, buried dozens of feet below them in a stage adapter ring, and it flew the rocket straight and true while they sorted out their problems.

Quickly the trio reset the breakers and got things back to normal. Conrad chuckled nervously. "That's one of the better sims, believe me," he said. Then, with a laugh, "Okay we're all chuckling up here over the lights, we all said there were so many on we couldn't read 'em..." More laughing followed.

Somehow, through the entire emergency, Pete Conrad did two remarkable things. First, he didn't twist the abort handle, on which he had maintained a firm grip. In a true emergency, he could rotate it, firing the escape system, whisking them away from the Saturn V and allowing them to parachute to safety—but he didn't do that. He waited out the problem.

But, perhaps more remarkably, the saltiest of the astronauts didn't curse once—after all, the world was listening, even President Nixon.

"Gol-durn almighty..." Conrad said, giggling. An event such as this was like tossing a match on Conrad's verbal gasoline puddle, but he still didn't curse. Instead, the crew began chattering and laughing about the incident as the rocket continued on into orbit. This small band of brothers had weathered their first glitch of the mission. It would not be the last.

Five days later, having reached lunar orbit and checked out their Lunar Module carefully, Conrad and Bean were headed down to the lunar surface while Gordon started his long vigil orbiting the moon alone. The Apollo 11 landing had been touch and go, and, for a variety of factors, Armstrong had overshot his intended landing site by miles, setting down with well under a minute's worth of fuel remaining. Conrad was determined not to do that, and he had worked with the mission planners to make sure he accomplished a pinpoint landing. They were aiming for a region in Oceanus Procellarum (the "Sea of Storms"), where NASA's robotic Surveyor 3 had landed in 1967. As Armstrong had done, Conrad let the computer do the flying for the early part of the descent. The tiny 36 kilobyte unit was enough to get them to the moon and back, although barely. But the computer had hiccupped during Armstrong's landing, a point both Conrad and his crewmate Bean were well aware of.

The area below them was called Mare Cognitum (the "Known Sea"), but the mission planners had taken to calling it Pete's Parking Lot.

The LM had been flying feetfirst, firing its single rocket engine to slow them down and allow them to fall toward the moon. Then they executed a command called "Pitch-Over," which put them vertical, windows forward.

Conrad practically squealed, "Hey, there it is! There it is! Son-of-a-gun! Right down the middle of the road!" He had sighted a crater called Snowman and knew they were close to their target. He took control of the LM and finessed the hand controller, making minute adjustments to bring them down. He continued to speak in a civil tongue, but in a technical debrief after his return, he said (in more characteristic verbiage) that at first it just looked like a black-and-white painting and he could not recognize anything. "I didn't know where the *fuck* I was! I looked out and I didn't know where the hell I was. I looked (at the computer) and got the number. I looked back through the number and *then* I knew where I was." The salty sailor was back.

At about 3,500 feet, Houston radioed up, "*Intrepid*, Houston. Go for landing." *Intrepid* was the call sign for the LM. Conrad responded, then went back to being a kid in a candy store. "That's so fantastic . . . I can't believe it!" Bean, ever the cool customer, responded, "You're at 2,000 feet . . ."

At 400 feet, Conrad took full control of the landing. Bean then got excited. "Hey! Look at that crater; right where it's supposed to be! Hey; you're beautiful. Ten percent . . ." referring to the amount of fuel remaining. It was time to land. "Come on down, Pete," Bean said.

Moments later, when an indicator light came on to let them know that they were within a few feet of the surface, Conrad cut the engine switch before the foot pads set down, as they had been trained to do. As the LM rapidly dropped the last few feet in the low lunar gravity (about 16 percent of Earth's), Conrad recalls starting to say, "Oh, shit—", but the thump of the landing cut him off.

They were on the moon, only the second manned craft to arrive. They were down to about the same amount of fuel Apollo 11 had remaining when they got there, but they were perfectly on target. The landing site was smooth, and they were within walking distance—about six hundred feet—of Surveyor 3, which they planned to visit. After they went through a sequence of switch throws to secure the LM for its stay on the moon, they got a call from Gordon orbiting overhead in the CM, *Yankee Clipper.*

"*Intrepid*, congratulations from *Yankee Clipper!*" Gordon said. Conrad responded, "Thank you sir, we'll see you in 32 hours."

Then Conrad looked out the window again, and cracked a grin. "Man, I can't wait to get outside, look at that." Bean chit-chatted a bit, and said, "Look at those boulders out there on the horizon, Pete. Jimmeny! This is a pretty good place. Look right over there." Conrad replied, somewhat dreamily, "Yeah."

About five hours later, they prepared to exit the LM and start their lunar exploration. It was time to test the Italian journalist's bet, and you better believe she was watching. Would Conrad actually say what he had made up on the spot during the press event just a few months earlier?

At 4 days, 19 hours, and 16 minutes mission elapsed time, Conrad started out the hatch, with his pal Bean guiding him, as Aldrin had done for Armstrong. Conrad paused at the top of the ladder and looked over to see what they called Surveyor Crater, about thirty feet away.

"Hey, I'll tell you what we're parked next to..." Conrad said. "What?" asked Bean. "We're about 25 feet in front of Surveyor Crater!" They both laughed.

Conrad reached the bottom rung and made the long drop to the lunar surface. The LM's legs were built to compress much more than they ever did upon landing, and the ladder had to be short enough not to crumple if the shock absorbing mechanism (crushable aluminum shock absorbers) lowered it further than expected. This meant that the top rung was a few feet above the surface. It had been a surprise to Armstrong, who was 5 feet, 11 inches tall. For the 5 foot, 6 inch Conrad, it felt a lot further. And so, when he set foot on the moon...

"Whoopie! Man, that may have been a small one for Neil, but it's a long one for me!" More laughter. He won the bet—but never did manage to collect on it.

Bean followed him down the ladder in short order, and for the next four hours they explored the moon. Conrad remained cognizant of his language while doing so, and instead of a steady barrage of off-color witticisms, took to humming to himself as he worked. Nothing tangible—you could not put a name to it, just "Dum-de-dum-da, dum de-de-dum."

Within minutes, Bean grabbed the color TV camera—the first on the moon (Apollo 11's had been low-resolution black and white), and started ambling toward a good spot to set it up. Unfortunately, they had not trained with the camera—it had not been ready, so they had used a block of wood instead—and Bean was not sufficiently careful with the lens. He paused for a second, swiveling the camera past the scorching sun, unmediated by an atmosphere. The video tube inside was a relatively new and miniaturized design, and extremely sensitive to harsh light. The sun burned the sensitive coating inside the video tube right off, and the camera was dead. Apollo 12 had just become a radio show.

**Figure 7.6. Al Bean descends the ladder of the Lunar Module. (Courtesy of NASA.)**

Bean was mortified, but there was work to be done; Conrad told him not to worry about it. They had a timeline to keep. They went about their chores, setting up experiments and collecting samples. Every now and then Conrad would just stop and laugh, and the controllers on the ground were not certain what he was thinking. His backup crew, however, knew perfectly well. Given the roustabout nature of the two moonwalkers, unlike the buttoned-down crew of Apollo 11, the other astronauts knew that these guys would enjoy a good joke. Dave Scott and Jim Irwin, their backups who would later fly on Apollo 15, had cut pictures out of *Playboy* magazine and taped them into the small checklists that Conrad and Bean wore on the cuffs of their spacesuits. Every few pages, Conrad would flip a card for a preview of the next bunch of tasks and find a compromising photo of Miss July, 1969. It cracked them up, as everything seemed to.

After setting up a US flag and the experimental package, and completing their rock and soil sample gathering, they headed back into the LM, tired and filthy, but elated. They closed up the LM, repressurized, and took off their suits. They were scheduled to eat and sleep before their second moonwalk, but trying to sleep while scrunched inside a tiny and very delicate spacecraft parked on the lunar surface seemed an unnatural act to both of them; Conrad drifted off into a light sleep while Bean fretted, finally getting some shuteye—until a cooling pump got noisy. They both awoke with a start, realized things were okay, and tried to go back to sleep. But neither of them were able to slumber again. And just to make things more annoying, Conrad's suit was adjusted wrong and was digging into his shoulder. They both got up to readjust it, which took the better part of an hour. Conrad finally called down to Houston and suggested they get their second EVA started early. Reluctantly, the doctors agreed, and the astronauts were given the go-ahead. It was now thirteen hours since they had ended their first activity outside the LM.

The second EVA would also last for just under four hours, and there was lots to do. Conrad and Bean explored a string of craters near the LM, God's very own geological drill holes—the older rock and soil inside the craters, which had been scattered by the impacts that had formed them, could be valuable for assessing the age and formation of the moon.

Soon the pair were far away from the LM, and both were getting winded. At one point, Conrad said, "I'll tell you what I'm going to do,

Houston. I'm going to take an EMU break. How you doing, Al?" Bean replied, "Okay." But this was not tech talk, an "EMU Break" was a code phrase. The EMU, or Extravehicular Mobility Unit, was a control box attached to the front of their suits that controlled the life-support back-packs. But there was no check needed; Conrad was letting Mission Control know that the two of them needed a momentary rest. As he said in a post-flight debrief, "We had sort of a code. They didn't want to say 'you guys look tired' and then the whole world says 'they're fucking fainting on the Moon.'" Conrad got his break, and then they were off again, huffing and puffing, while Conrad hummed his way through the procedures.

One of the more remarkable parts of this second outing was their excursion to Surveyor 3. The robotic probe had landed on the moon in 1967, and the scientists wanted to see what the effects of two years of exposure to the lunar environment might be. The two men trotted the six hundred feet to the machine, Conrad wielding a pair of bolt cutters to snip the camera off of the unsuspecting robot. With a bit of diffi-culty, he did just that, and put the spindly artifact into a bag to bring it back to Earth.

The two explorers wrapped up their workday and headed back to the LM. They would soon be departing the lunar surface, and there was still plenty to do before they left. They had to stow all their col-lected samples, toss out unneeded items to save weight, and prepare the LM for liftoff. Then Bean started obsessively checking the settings for the ascent coming up in a couple of hours. Conrad noticed him going over things multiple times, and said, "Beano, are you worried about the engine?" to which Al replied with a simple, "Yeah." Conrad chuckled and told him not to worry, it would work; and if it didn't— "we're just gonna become the first permanent monument to the space program!" Al didn't laugh at that one.

But work it did, and within hours they were on their way back to their pal Gordon. The two spacecraft completed their lunar rendez-vous, and Gordon opened the hatch. He was thrilled to see his crew-mates safe and sound, but they were beyond scruffy, even beyond dirty . . . they were filthy.

**Figure 7.7. Conrad holds the Surveyor 3 TV camera, which he took back to the LM after cutting it loose from the robotic probe. (Courtesy of NASA.)**

He admonished them to keep that dirt out of his nice clean Command Module. In truth, he may have been somewhat concerned about the effects of all that lunar dust and dirt on the switches and air recirculation systems inside—a bit of grit drifting into the wrong place could cause real problems. "You guys ain't gonna mess up my nice clean spacecraft!" he yelled down the docking tunnel. So, ever the gentleman, Conrad decided to accommodate him. A short time later, he and Bean drifted up into Gordon's capsule just as God had made them—without a stitch of clothing. Who was laughing now?

"So picture this," Conrad said, as he stretched out in his God-given glory, "Something really bad happens, right now, and we're done. And somebody finds us a thousand years from now, just like this. What do they conclude?"[14]

"That I am a sick and lonely man," Gordon retorted, "and I went through a lot of trouble and expense for some privacy."

The trip home was uneventful, but full of chatter. You could just not shut these guys up. But Houston put up with it—the mission had been another in a string of successes, despite the burned-out camera and other small inconveniences. The trio splashed down in the warm waters of the Pacific Ocean ten days after they had left Earth. The astronauts were retrieved and flown first to Pago Pago, then Hawaii, on their way back to Houston. Apollo 12 headed off into the history books.

But their ordeal was not quite over. There will still concerns about biological contaminants returning from the moon, so they were suited up in biocontainment suits upon exiting the capsule, and kept in an isolation unit—the same converted, airtight Airstream trailer that had housed the Apollo 11 crew—until they were transferred to larger digs. They spent a total of three weeks like this, until they were finally unleashed upon the world, none the worse for having cavorted on the moon. After a whirlwind of press and NASA public relations activity (not even close to what the Apollo 11 crew had endured, but still grueling), the three of them headed back into preparations for what lay ahead.

Gordon stayed in training for what he hoped would be another Apollo mission. He never got it—the last three Apollo flights had been canceled, and he entered the private sector in 1972.

Conrad and Bean shifted over to Skylab, the only project that made it out of Bean's old posting at the Apollo Applications Project. By utilizing the hardware remaining from the cancellation of Apollos 18, 19, and 20, NASA was able to convert the third stage of the Saturn V into a space station. Skylab was built from surplus rocket parts, but that in no way detracted from the beauty of the project. It was going to be spectacular. Conrad had been eyeing Skylab since serious consideration of the mission started in 1968, and he wanted in. As so often happened, he got his way, becoming the commander of the first crew to visit the space station. Bean followed him, and commanded the second crew to Skylab.

Skylab was almost 83 feet long and 22 feet in diameter. After living in the Apollo Command Module for the better part of two weeks, and prior to that inside the tiny Gemini capsule for one week, Skylab would be the Hilton of space for Conrad. With 12,750 cubic feet of interior space, it was just enormous. And that interior had been thoughtfully laid out by a program on a budget. Although funds had been tight since the budget cuts following the Apollo program, NASA had sought out the services of famed industrial designer Raymond Loewy, who had created streamline locomotives, the original Coca-Cola vending machines, and some of the most famous logos in the world (Shell, Exxon, and TWA, for example). Loewy designed Skylab's interior to be as attractive and crew-friendly as a purpose-built space station could be. Skylab would be the best career-capping mission Conrad could have asked for.

Skylab launched on May 14, 1973, and flew right into trouble. During launch, the two enormous wing-like solar panels folded back on the sides of the space station were damaged. A micrometeoroid shield, a thin shroud of sheet metal, had torn free. On one side, the solar panel had gone with it, fluttering down into the Atlantic Ocean. On the other side, the solar panel was still there, but it was stuck fast against the side of the station. Without these, Skylab had only a much smaller solar panel assembly mounted on the front to power it, and it provided just a small trickle of energy. Without more electricity, the onboard cooling systems would not function, and, bathed in blistering sunlight, Skylab would slowly roast into a hunk of junk in orbit.

Pete Conrad, Paul Weitz, and Joe Kerwin were the first crew slated to inhabit the space station. They had been training for this since shortly after Conrad's Apollo mission, and they were supposed to launch a day after Skylab, fly to it, float inside, and set up shop for a month. But the three crewmen had been forced to stand down and watch as Mission Control tried to figure out what all the wonky power and temperature readings from Skylab were trying to tell them. The engineers soon realized what had happened and started devising solutions and fixes. On May 25, Conrad and crew, hurriedly trained in repair procedures, finally set out to rendezvous with Skylab and get to work.

As Conrad closed on the crippled station, it was quickly apparent what had gone wrong. He had been intimately involved with the design and construction of Skylab and knew it as only a hands-on engineer could. But up till now, he'd only had the indirect deductions of the NASA engineers and some fuzzy telescopic photos to work with. Now he could see Skylab up close—and it was a mess.

Regardless of the visible damage, Conrad kept his enthusiasm high. "Skylab, tally-ho!" he said exuberantly, as he, Weitz, and Kerwin closed on the nearly derelict station. They had a lot of work ahead of them, but at least they were there. Conrad was not known for his patience and had been crawling out of his skin to get launched.

Circling the station in the Apollo capsule, Conrad assessed the damage. He verified that one solar panel was gone, and the other was tangled up in the damaged metal micrometeorite shield.

The astronauts flew around Skylab, describing what they saw, along with live TV images. They docked with the station temporarily, then an hour later disengaged to try to see what they might be able to do to get the remaining solar panel deployed. Somehow they would have to clear the sheet metal wreckage from the panel and hope that the hinge was still intact, so that the panel could swing free and lock as designed. Conrad flew the spacecraft to a point near the panel, and Weitz opened the hatch of the CM and stood up, while Kerwin held his legs, to try and wrest the entanglement free. As Weitz tugged and pulled, Conrad did his best to keep the Apollo spacecraft from banging into the side of Skylab, but it was touch and go, and, he felt, dangerous. Despite his reputation as a gung-ho astronaut, he knew that any damage to the Apollo spacecraft could end the mission on the spot. With little to show for the effort, the frustrated crew redocked twenty-seven hours after launching—it took eight tries to do so, but they were finally able to reenter the overheating station and settle in for a rest despite the high temperatures inside. They were exhausted.

The next day, the astronauts were able to do a partial repair by setting up a large cover over part of the station, which lowered temperatures somewhat—they called it a parasol, and it was deployed through a small airlock using a remote-control mechanism. But the station needed that stuck solar panel to be functional to complete their mission.

Conrad, Weitz, and Kerwin spent the next week and a half operating the station at reduced power, making the best of the marginal conditions while NASA figured out what to do. Mission Control had uplinked procedural instructions on how they *thought* the task could be achieved, and the resulting printout was fifteen feet long—a hell of a lot of instructions. After studying the list in detail, Conrad and Kerwin suited up and left Skylab through a hatch to attempt the repairs.

While the procedures had been thoroughly rehearsed in the neutral buoyancy tank on Earth, actually climbing around the outside of Skylab was more challenging than had been envisioned. It was like the frustrations during the EVAs on the Gemini flights—there were too few handholds and gripping points on the exterior; this kind of repair had never been anticipated. But the two astronauts made do and managed to reach the stricken solar panel.

They carried with them a long pole with a cutting tool on the end—a modified tree-branch cutter that NASA had acquired from the Southwestern Bell telephone company. After much struggle, Conrad and Kerwin managed to get the pole's head attached to an aluminum strap that was snagging the far end of the panel. Conrad then used the pole to work his way out to the end, hand over hand. He tied a rope he had trailed behind him to the far point, then returned to Kerwin, who had been working to find a place to tie off the other end on the station's slippery exterior. At that moment, the cutters on the end of the pole jerked, cutting through the aluminum strip, and the metal shroud popped off, releasing the panel—and Conrad. The pole drifted free, as did Conrad. It was a potential nightmare, with Conrad drifting around near a piece of sharp sheet metal. Conrad could get tangled in his tether, the sheet metal could cut his suit, and the mission could end with a dead astronaut hanging from the end of his umbilical. But, true to form, Conrad laughed it off and got back to Skylab just fine—holding his tongue for the most part, other than an exuberant, "Whooo-yaaa! We got it!" as the two pulled themselves back to the space station.[15]

The solar panel was now partially deployed, but it was still stuck partway and not generating current. But with the rope tied off at both ends, the solution was in sight. Without detailed consultation with Houston, Conrad instructed Kerwin to join him in crouching underneath the rope, holding it above their bulky suits. On cue, both stood up, pulling the rope away from Skylab and yanking the solar panel free. It swung open, locking in the 90-degree position it had been designed to do. The rope twanged as the panel locked, and once again Conrad, now joined by Kerwin, cartwheeled into space, dangling from their umbilicals.

As Conrad put it when talking to Mission Control after they reentered Skylab, it had been quite a moment.

The CAPCOM in Houston initiated the commentary: "While I've got you here, we did have a question on the way the SAS panels came out. We'd like to know whether they jumped out to about where they ended up or did they jump out and then ease on out?"

Conrad laughed heartily. "I'm sorry you asked that question. I was facing away from it, heaving with all my might. And Joe was also heaving with all his might when it let go. And both of us took off. And by the time we got ourselves under control and back down and around the spacecraft some place again, lo and behold, the SAS panel [the technical name for the solar panel, the Solar Array System] was already out and locked. So I can't answer that question for you. By the time we got settled down and looked at it, those panels were out as far as they were going to go at the time." He did not ask about how many "EXPLETIVES DELETED" would be required in the official transcript of the radio chatter during the repair, and Mission Control didn't tell him.

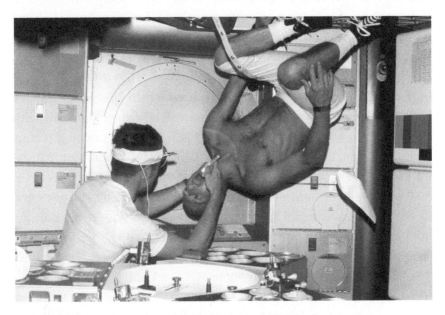

Figure 7.8. Conrad has his teeth examined, gaps and all, by Joe Kerwin. (Courtesy of NASA.)

The temperature inside Skylab dropped to normal levels, and the increase in power from the large remaining solar panel allowed them to get on with the objectives planned for the final two weeks of the mission.

Skylab was huge and, with enough power to make it work, relatively comfortable. The crew was now able to conduct overdue experiments, exercise, eat, and sleep as they had been intended to. In the movies the astronauts took onboard, Conrad can be seen running around the inside of the cylindrical hull, laughing and tumbling as he drifted through the interior. It looked like fun and games, but a lot of good science resulted from their 392 hours of experimental procedures. They performed many medical experiments, returned 29,000 photos taken of the sun from Skylab's solar telescope, and spent a total of twenty-eight days in space, double the previous US record.

**Figure 7.9.  Artist's impression of McDonnell Douglas's DC-X "Delta Clipper." (Courtesy of NASA/MSFC.)**

Conrad felt that, despite the momentous nature of the Apollo moon landings, Skylab was a career high for him. He retired from NASA and the navy shortly after returning from the Skylab mission and took a job as a vice president of the American Television and Communications Company. Three years later, he moved on to McDonnell Douglas, a giant aerospace contractor, working with their commercial aircraft division. In the 1990s he started work on the company's DC-X Delta Clipper, an experimental single-stage-to-orbit (SSTO) spacecraft, which took off like rocket, then returned to Earth the same way, landing legs down. While ultimately mothballed before any commercial use could be made of it, the Delta Clipper pioneered some of the technologies now in use by SpaceX for their Falcon 9 reusable rockets.

Conrad was early to the table with his notions of commercializing spaceflight. "It's not until we get into the commercial world where space begins to pay for itself that you're going to see things like that," he responded in a PBS television interview asking him about the possible flight of everyday people into space and to the moon.[16] "Everybody forgets old Christopher Columbus didn't sail across the Atlantic for the good of all mankind," he continued. "He sailed across to make a buck, and he was supported by a government that wanted to make a buck. . . . We've got to get the commercial world really going and bring those costs of getting into orbit down." Sadly, he would not live to see this begin to reach fruition in the second decade of the twenty-first century.

Conrad died on July 8, 1999, as the result of a motorcycle accident. Even in his later life, he loved things that moved fast, and motorcycles remained a passion. At the time of the crash, he was riding at a moderate speed near Ojai, California, when he lost control of his bike. His wife, Nancy, who was riding another motorcycle, came upon him and stayed at his side until emergency crews took him to the hospital, where he succumbed to his injuries. She was heartbroken, but she knew he had lived a full life.

As Nancy Conrad had said earlier, in a 1996 *Los Angeles Times* interview, "What this guy has done in 10 minutes on any given day is more than most people do in 10 years. . . . He has a lot of pies in front of him, and he's sticking his finger in all of them to see if he can pull out a plum. He's a true explorer."[17]

At Conrad's memorial service, Neil Armstrong, barely controlling his rarely seen emotions, added, "I'm not sure what he's doing right now but I suspect he's telling some stories of the old days . . . Pete was the best man I ever knew. He treated me like a brother."[18]

Conrad's tombstone sums it up. Below his name and dates of his life, his epitaph says, simply, "An Original." And indeed he was.

# CHAPTER 8

# HEROES OF A NEW SPACE AGE

In these pages, you have met just the smallest sampling of people who were a seminal part of the first human expansion beyond our planet. They are just some of the amazing people who took those first fledgling steps into the last and greatest frontier, a vast realm beyond our world that offers an endless bounty for humanity—both in space and here on Earth. The next great expansion of our species is coming, and coming soon, for we are on the cusp of a new age of exploration in space.

As you read this, in countless places all over the world a new generation of engineers, scientists, businessmen, investors, tinkerers, and technicians are working tirelessly to take this next great step. Spaceflight is no longer the exclusive domain of two superpowers as it was in the 1960s, and the driving force is no longer the great game of geopolitics that was at the core of the first space age. For reasons ranging from the imperiling of our home planet by overpopulation and rampant industrialism to the emerging business opportunities offered by the development of space itself as a resource, new efforts are being undertaken in places and in ways that could never have been imagined during the first space age. To be sure, powerful and inspiring programs will continue to be driven by governments, which now include not just the United States and Russia, but increasingly China, Japan, India, and a number of European countries. But participation in spaceflight by private industry, driven both by visionary billionaires and regular citizens, is emerging in most developed countries, and is slowly but inexorably spreading to smaller, less industrialized nations as well. With much of the heavy lifting done by the competition to land a man on the moon in the twentieth century—a process that defined and developed the root technologies that can carry humans to Earth orbit and beyond—enterprises both large and small are utilizing new technologies born of the twenty-first century to take the next steps in the development of the solar system for the benefit of all humanity.

The most obvious examples of these enterprises are the tireless efforts by American billionaires and entrepreneurs Elon Musk and Jeff Bezos in the United States. Musk's SpaceX is building new and powerful rockets that have captured much of the global launch market, and is currently fabricating the largest rocket in history, the Starship, that will be able to carry enormous cargoes and large groups of passengers to the moon and Mars. For his part, Bezos is investing a billion dollars of his personal wealth each year into his own rocket company, Blue Origin, to design and build rockets that will perform tasks ranging from the carrying of well-heeled tourists on suborbital excursions, to boosters that will launch huge payloads into space, and landers that will carry payloads to the moon and back, enabling ever greater adventures. Both companies will perform these feats at prices that will allow a vast acceleration of space enterprise—it is truly a new era in spaceflight.

Smaller undertakings follow directly behind them. Richard Branson's Virgin Galactic is poised to carry space tourists on short trips to near orbit, as well as to deliver satellites and other space cargo into orbit at unheard of prices with ever increasing frequency. Rocketlab, based in both the United States and New Zealand, is flying smaller rockets that will carry the increasingly miniaturized payloads that will soon dominate our orbital operations. Other companies in the United States, Russia, and China are endeavoring to follow suit and are making great strides.

But rockets are just one facet of the new age of spaceflight. Other companies, mostly small and funded by private investment, are creating 3-D printers that will be able to use raw materials carried into orbit, and, later, resources sourced from asteroids and the moon, to print the machines and structures that will open the final frontier.

New software firms are popping up globally that utilize data derived from satellites to refine and streamline human enterprise, from agriculture to freight transport to retail sales (Target and Walmart use satellite imaging to count shoppers in their competitor's parking lots, for example). New investment firms have come to the fore to channel money from investors all over the world into space-related enterprises, and this investment extends into double-digit billions each year and is growing.

And this trend extends beyond moneymaking enterprises. In the spirit initiated by the American SETI Institute, Yuri Milner, a Russian billionaire, has pledged $100 million of his money to discovering possible intelligent extraterrestrial species with his Breakthrough initiatives—a purely scientific endeavor. Other similar science- and exploration-driven efforts are likely to follow.

In university labs and garage workshops worldwide, students and young entrepreneurs are building a new generation of tiny spacecraft that will perform a wide array of functions in low Earth orbit and beyond—they are capitalizing on the fact that microprocessors such as the ones found in your cellphone can now do the same work that warehouses full of computers did in the 1960s. There are even hackathons where large groups of young programmers assemble to code the future with no expectation of financial return.

Into this new marketplace steps the youth of today—educated, driven, and impatient—who will continue to open the space frontier for their generation. They come from many nations and backgrounds, with a common goal of bringing the bounty to be derived from the development of space back to their home countries.

All these endeavors rely on new and engaged minds to drive the expansion of humanity into space. Hundreds of thousands of people from all walks of life will be needed, and, increasingly, there are high-paying jobs in the space sector to employ them. From these ranks will emerge the new icons of spaceflight. Some will become famous and wealthy—studies indicate that the first trillionaires are likely to emerge from the space business sector—and others will labor in relative anonymity, satisfied with the knowledge that they are engaged in the work that drives them and inflames their passions.

These will be the leaders of the new space age, whose brilliance and dedication will transform the lives of each and every one of us.

Here's to the new heroes of spaceflight.

# NOTES

## CHAPTER 1. YURI GAGARIN: THE FIRST STARMAN

1. Boris Alexeev, *Unified Non-Local Theory of Transport Processes: Generalized Boltzmann Physical Kinetics* (Amsterdam: Elsevier, 2004), p. 164.

2. Diana Falzone, "7 Hollywood Stars Who Are Completely Out of Touch with Reality," Fox News, February 1, 2017, https://www.foxnews.com/entertainment/7-hollywood-stars-who-are-completely-out-of-touch-with-reality.

3. Francis French and Colin Burgess, *Into That Silent Sea: Trailblazers of the Space Era, 1961–1965* (Lincoln: University of Nebraska Press, 2007), p. 2.

4. Ibid., p. 11.

5. "The Flight of Vostok 1," *50 Years of Humans in Space*, European Space Agency, March 21, 2011, https://www.esa.int/About_Us/Welcome_to_ESA/ESA_history/50_years_of_humans_in_space/The_flight_of_Vostok_1.

6. Ibid.

7. Ibid.

8. French and Burgess, *Into That Silent Sea*, p. 24.

9. Colin Stuart, *How to Live in Space: Everything You Need to Know for the Not-So-Distant Future* (Washington, DC: Smithsonian Books, 2018).

10. Stuart Williams, "Russia Marks 50 Years Since Gagarin Triumph," Phys.org, April 7, 2011, https://phys.org/news/2011-04-russia-years-gagarin-triumph.html.

11. French and Burgess, *Into That Silent Sea*, p. 26.

12. Williams, "Russia Marks 50 Years Since Gagarin Triumph."

13. While the term "UFO" refers to any flying object that is unidentified, in this case the conjecture was that some alien entity was at work.

## CHAPTER 2. JOHN GLENN: THE CLEAN MARINE

1. John Glenn, *John Glenn: A Memoir*, with Nick Taylor (New York: Bantam Books, 1999), p. 135.

2. Ibid., p. 5.

3. Ibid., p. 11.

4. Ibid.

5. Ibid., p. 36.

6. Ibid., p. 37.

7. Ibid., p. 135.

8. Ibid., p. 173.

9. Ibid.

10. Ibid., pp. 233–344.

11. Glenn, *John Glenn*, p. 179.

12. Glenn, interview by Thomas, April 21, 2008, pp. 4–5.

13. Ibid.

14. Glenn, *John Glenn*, pp. 182–83.

15. Ibid., pp. 183, 184.

16. Glenn, interview by Thomas, April 21, 2008, p. 7.

17. Ibid., pp. 7–8.

18. Ibid., pp. 8–9.

19. Samples of MMPI questions from https://antipolygraph.org/cgi-bin/forums/YaBB.pl?num=1109032158.

20. Glenn, *John Glenn*, p. 189.

21. Glenn, interview by Thomas, April 21, 2008, pp. 18, 19.

22. "Press Conference: Mercury Astronaut Team," transcript, NASA, April 9, 1959, https://www.nasa.gov/pdf/147556main_presscon.pdf.

23. Ibid.

24. Ibid.

25. Glenn, interview by Thomas, April 21, 2008, p. 29.

26. Ibid., pp. 29–30.

27. Ibid., pp. 31–32.

28. Ibid., p. 32.

29. John F. Kennedy, "Address to Joint Session of Congress, May 25, 1961," *John F. Kennedy Presidential Library and Museum*, https://www.jfklibrary.org/learn/about-jfk/historic-speeches/address-to-joint-session-of-congress-may-25-1961.

30. Glenn, *John Glenn.*

31. Ibid., p. 258.

32. Mercury 7 Archives, NASA Kennedy Space Center Telescience and Internet Systems Lab, https://science.ksc.nasa.gov/history/mercury/ma-6/sounds/.

33. John H. Glenn Jr., interview by Jeffrey W. Thomas, John Glenn Archives, Ohio State University, May 23, 2008, p. 22, https://kb.osu.edu/bitstream/handle/1811/79328/OCA-JohnGlenn-Session19-transcript.pdf?sequence=1&isAllowed=y.

34. John Saavedra, "The Mystery of the 'Fireflies' That Swarmed John Glenn's Spaceship," *The Space Page*, https://www.ranker.com/list/john-glenn-orbital-firefly-mystery/john-saavedra.

35. Ibid.

36. Glenn, interview by Thomas, May 23, 2008, p. 31.

37. Ibid., p. 29.

38. Associated Press. "Apollo 11: How America Won the Race to the Moon." Diversion Books, 2016.

39. Ibid., p. 43.

40. Glenn, *John Glenn*, p. 284.

41. Ibid., p. 472.

42. Tara Gray, "John H. Glenn, Jr.," *40th Anniversary of the Mercury 7*, NASA History Office, https://history.nasa.gov/40thmerc7/glenn.htm.

43. John Glenn, *John Glenn: A Memoir*, with Nick Taylor (New York: Bantam Books, 1999).

44. Julie Zauzmer, "In Space, John Glenn Saw the Face of God: 'It Just Strengthens My Faith," *Washington Post*, December 8, 2016, https://www.washingtonpost.com/news/acts-of-faith/wp/2016/12/08/in-outer-space-john-glenn-saw-the-face-of-god/?utm_term=.62f20e4f6a91.

## CHAPTER 3. VALENTINA TERESHKOVA: FLIGHT OF THE SEAGULL

1. Valentina Tereshkova, as quoted in *Into That Silent Sea: Trailblazers of the Space Era, 1961–1965*, by Francis French and Colin Burgess (Lincoln: University of Nebraska Press, 2007), p. 298.

2. Valentina Tereshkova, *The First Lady of Space: In Her Own Words* (Bethesda, MD: SpaceHistory101.com Press, 2015), p. 23.

3. French and Burgess, *Into That Silent Sea*, p. 317.

4. Tereshkova, *First Lady of Space*, pp. 30–31.

5. Mark Wade, "Tereshkova, Valentina Vladimirovna," *Encyclopedia Astronautica*, http://www.astronautix.com/t/tereshkova.html.

6. James E. Oberg, *Red Star in Orbit: The Inside Story of Soviet Failures and Triumphs in Space* (New York: Random House, 1981), p. 69.

7. Tereshkova, *First Lady of Space*, p. 40.

8. Ibid., p. 41.

9. Oberg, *Red Star in Orbit*, p. 68.

10. Tereshkova, *First Lady of Space*, p. 42.

11. Oberg, *Red Star in Orbit*, p. 69.

12. James Oberg, "Does Mars Need Women? Russians Say No," NBC News, February 11, 2005, http://www.nbcnews.com/id/6955149/ns/technology_and_science-space/t/does-mars-need-women-russians-say-no/#.XDYZu1VKhpg.

13. Ben Evans, *Escaping the Bonds of Earth: The Fifties and the Sixties* (Berlin: Springer, 2010), p. 57.

14. Robin McKie, "Valentina Tereshkova, 76, First Woman in Space, Seeks One-Way Ticket to Mars," *Guardian*, September 17, 2013, https://www.theguardian.com/science/2013/sep/17/mars-one-way-ticket.

15. Tereshkova, *First Lady of Space*, p. 10.

## CHAPTER 4. GENE KRANZ: STARS AND STRIPES FOREVER

1. Colin Pesyna, "Lessons in Manliness from Gene Kranz," guest post on *The Art of Manliness* (blog), by Brett and Kate McKay, July 20, 2009, https://www.artofmanliness.com/articles/lessons-in-manliness-from-gene-kranz/.

2. Gene Kranz, *Failure Is Not an Option: Mission Control from Mercury to Apollo 13 and Beyond* (New York: Simon & Schuster, 2001), pp. 102–103.

3. "Gene Kranz Talks about His High School Term Paper," YouTube video, 1:23, originally aired on *InnerVIEWS with Ernie Manouse*, posted by douglas martin, September 11, 2016, https://www.youtube.com/watch?time_continue=83&v=jFk0gNTIpbw.

4. Kranz, *Failure Is Not an Option*, p. 20.

5. Ibid., p. 21.

6. Ibid., pp. 234–35.

7. Ibid., p. 38.

8. Ibid., pp. 56–57.

9. Ibid., p. 59.

10. Ibid., p. 69.

11. Ibid., p. 73.

12. Ibid., p. 92.

13. Ibid., p. 147.

14. Ibid., p. 187.

15. Eugene F. Kranz, interview by Rebecca Wright, Carol Butler, and Sasha Tarrant, Johnson Space Center Oral History Project, January 8, 1999, in *Before This Decade Is Out . . . : Personal Reflections on the Apollo Program*, ed. Glen E. Swanson (Washington, DC: NASA History Office, 1999), p. 123.

16. Ibid., p. 124.

17. Ibid., pp. 124–25.

18. Pesyna, "Lessons in Manliness from Gene Kranz."

19. Michael Cabbage, "40 Years Later, Recalling the Lessons of Apollo 1," *Los Angeles Times*, January 28, 2007, http://articles.latimes.com/2007/jan/28/nation/na-apollo28.

20. Kranz, interview by Wright, Butler, and Tarrant, p. 131.

21. Andrew Chaikin, *A Man on the Moon: The Voyages of the Apollo Astronauts* (New York: Penguin Books, 1998), p. 280.

22. Ibid., pp. 134–35.

23. Ibid., p. 136.

24. Ibid., pp. 136–37.

25. Ibid., p. 137.

26. Ibid., p. 139.

27. Ibid., p. 141.

28. Ibid., pp. 141–42.

29. Ibid., p. 142.

30. Ibid., p. 144.

31. Ibid.

32. James R. Hansen, *First Man: The Life of Neil Armstrong* (New York: Simon & Schuster, 2005), p. 463.

33. Kranz, interview by Wright, Butler, and Tarrant, pp. 147–48.

34. Ibid., p. 151.

35. Ibid., p. 148.

36. Ibid., p. 151.

37. Gene Kranz, interview by the author, October 30, 2005.

38. Kranz, *Failure Is Not an Option*, pp. 283–84.

39. Kranz, interview by the author.

40. Kranz, interview by Wright, Butler, and Tarrant, p. 155.

41. Ibid., pp. 155–56.

42. "Apollo 11 Lunar Descent Flight Director's Loop HD," YouTube video, 16:25, posted by Austin1987VCR, August 10, 2014, https://www.youtube.com/watch?v=QKdKBILTUK4&t=7s, at 6:09.

43. Kranz, *Failure Is Not an Option*, p. 284.

44. Kranz, interview by the author.

45. Kranz, interview by Wright, Butler, and Tarrant, p. 157.

46. Eric M. Jones, ed., "The First Lunar Landing," *Apollo 11 Lunar Surface Journal*, NASA Historical Archives, 1995, last revised May 10, 2018, https://www.hq.nasa.gov/alsj/a11/a11.landing.html.

47. Kranz, interview by Wright, Butler, and Tarrant, p. 159.

48. Kranz, *Failure Is Not an Option*, p. 289.

49. Kranz, interview by Wright, Butler, and Tarrant, pp. 160–61.

50. Charlie Duke, interview by the author, November 2, 2005.

51. Kranz, interview by Wright, Butler, and Tarrant, p. 161.

52. Ibid., pp. 162–63.

53. Ibid., p. 163.

54. Gene Kranz, interview by Roy Neal, Johnson Space Center Oral History Project, April 28, 1999, available online at https://www.c-span.org/video/?292341-2/gene-kranz-oral-history-interview-part-2&start=494, 8:23–9:49.

55. Ibid., 14:49–14:52, 16:00–16:04.

56. Ibid., 16:40–17:31.

57. Ibid., 18:00, 19:24–19:35, 21:50–22:10.

58. Ibid., 22:43–23:03.

59. Ibid., 24:25–25:00, 25:09–25:18.

60. Ibid., 25:29–26:23.

61. Ibid., 34:26–35:38.

62. Ibid., 38:39–39:08.

63. Ibid., 39:08–39:35.

64. Ibid., 40:10–40:28.

65. Ibid., 43:04–43:25.

66. Ibid., 43:48–44:20.

67. Ibid., 45:47–46:17.

68. Ibid., 48:43–48:54.

69. Ibid., 57:48–58:04.

70. Ibid., 58:05–59:15.

## CHAPTER 5. MARGARET HAMILTON: THE FIRST SOFTWARE ENGINEER

1. Eric M. Jones, ed., "The First Lunar Landing," *Apollo 11 Lunar Surface Journal*, NASA Historical Archives, 1995, last revised May 10, 2018, https://www.hq.nasa.gov/alsj/a11/a11.landing.html, 102:28:08.

2. Gene Kranz, interview by the author, March 2005.

3. Ibid.

4. David Leonhardt, "John Tukey, 85, Statistician; Coined the Word 'Software,'" *New York Times*, July 28, 2000, https://www.nytimes.com/2000/07/28/us/john-tukey-85-statistician-coined-the-word-software.html.

5. "Margaret Hamilton, NASA's First Software Engineer," *Makers*, https://www.makers.com/profiles/596e0f42bea17725160a95c1/596d12 168c08e024562f9b9b, 00:55–1:16.

6. Hans Dieter Hellige, "Actors, Visions, and Developments in the History of Computer Communications" (paper presented at the Symposium "Technohistory of Electrical Information Technology," Munich, Germany, December 1990).

7. Margaret Hamilton, Apollo Guidance Computer History Project, First Conference, July 27, 2001, https://authors.library.caltech.edu/5456/1/hrst.mit.edu/hrs/apollo/public/conference1/hamilton-intro.htm.

8. Lori M. Cameron, "What to Know about the Scientist Who Invented the Term 'Software Engineering,'" *IEEE Software*, June 8, 2018, https://publications.computer.org/software-magazine/2018/06/08/margaret-hamilton-software-engineering-pioneer-apollo-11/.

9. Ibid.

10. Hamilton, Apollo Guidance Computer History Project.

11. Adam Fabio, "Margaret Hamilton Takes Software Engineering to the Moon and Beyond" *Hackaday*, April 10, 2008, https://hackaday

.com/2018/04/10/margaret-hamilton-takes-software-engineering-to-the
-moon-and-beyond/.

12.  Robert McMillan, "Her Code Got Humans on the Moon—and Invented Software Itself," *WIRED*, October 13, 2015, https://www.wired
.com/2015/10/margaret-hamilton-nasa-apollo/.

13.  Fabio, "Margaret Hamilton."

14.  Hamilton, Apollo Guidance Computer History Project.

15.  Fabio, "Margaret Hamilton."

16.  Dag Spicer, "2017 CHM Fellow Margaret Hamilton," video, Computer History Museum, April 27, 2017, https://www.computerhistory
.org/atchm/2017-chm-fellow-margaret-hamilton/, 4:25–4:48.

17.  McMillan, "Her Code Got Humans on the Moon."

18.  Spicer, "2017 CHM Fellow Margaret Hamilton," 5:16–6:02.

## CHAPTER 6. NEIL ARMSTRONG AND BUZZ ALDRIN: "FIRST MEN"

1.  Neil A. Armstrong, interview by Stephen Ambrose and Douglas Brinkley, NASA Johnson Space Center Oral History Project, September 19, 2001, https://www.nasa.gov/pdf/62281main_armstrong_oralhistory.pdf, p. 3.

2.  Ibid., p. 4.

3.  Ibid., p. 8.

4.  Ibid., pp. 17–18.

5.  Ibid., pp. 26–27.

6.  Ibid., p. 27.

7.  Ibid.

8.  Ibid., pp. 32–33.

9.  Ibid., p. 33.

10.  John F. Kennedy, "Excerpt from the 'Special Message to the Congress on Urgent National Needs,'" May 25, 1961, NASA History, May 24, 2004, https://www.nasa.gov/vision/space/features/jfk_speech_text.html.

11.  James R. Hansen, *First Man: The Life of Neil A. Armstrong* (New York: Simon & Schuster, 2005), p. 195.

12.  Armstrong, interview by Ambrose and Brinkley, p. 39.

13.  Ibid., pp. 39–40.

14.  Hansen, *First Man*, p. 221.

15. Buzz Aldrin and Malcolm McConnell, *Men from Earth* (New York: Bantam Books, 1989), p. 69.

16. Armstrong, interview by Ambrose and Brinkley, p. 43.

17. Ibid.

18. Edwin "Buzz" Aldrin, interview by Robert Merrifield, NASA Johnson Space Center Oral History Project, University of Houston–Clear Lake, July 7, 1970, https://uhcl-ir.tdl.org/handle/10657.1/847, p. 3.

19. Armstrong, interview by Ambrose and Brinkley, p. 53.

20. This and subsequent air-to-ground transmissions: "Gemini VIII Composite Air-to-Ground and Onboard Voice Tape Transcription," NASA Johnson Space Center History Portal, https://www.jsc.nasa.gov/history/mission_trans/gemini8.htm.

21. Armstrong, interview by Ambrose and Brinkley, pp. 55–56.

22. Ibid., p. 53.

23. Ibid., p. 54.

24. "First Man on the Moon," *NOVA*, season 41, episode 23, directed by Duncan Copp and Christopher Riley, aired November 29, 2014, on PBS, https://www.pbs.org/wgbh/nova/video/first-man-on-the-moon/.

25. Aldrin and McConnell, *Men from Earth*, pp. 153–54.

26. Ibid., p. 157.

27. Ibid., p. 159.

28. Armstrong, interview by Ambrose and Brinkley, p. 59.

29. Aldrin and McConnell, *Men from Earth*, p. 165.

30. Ibid., p. 166.

31. Armstrong, interview by Ambrose and Brinkley, p. 60.

32. Ibid., p. 61.

33. Ibid., p. 44.

34. Ibid., p. 69.

35. Ibid.

36. Ibid., p. 70.

37. Ibid.

38. Ibid., p. 62.

39. Ibid., p. 64.

40. Ibid., pp. 70–71.

41. Hansen, *First Man*, p. 392.

42. Ibid., p. 393.

43. Armstrong, interview by Ambrose and Brinkley, p. 81.

44. Ibid.

45. Hansen, *First Man*, pp. 396–97.

46. Buzz Aldrin, *Magnificent Desolation: The Long Journey Home from the Moon*, with Ken Abraham (London: Bloomsbury, 2009), p. 16.

47. W. David Woods, Kenneth D. MacTaggart, and Frank O'Brien, eds., "Day 1: Launch," *Apollo 11 Flight Journal*, https://history.nasa.gov/afj/ap11fj/01launch.html.

48. W. David Woods, Kenneth D. MacTaggart, and Frank O'Brien, eds., "Day 1: Transposition, Docking, and Extraction," *Apollo 11 Flight Journal*, https://history.nasa.gov/afj/ap11fj/03tde.html, 004:52:19.

49. Aldrin and McConnell, *Men from Earth*, p. 233.

50. W. David Woods, Kenneth D. MacTaggart, and Frank O'Brien, eds., "Day 4: Entering Lunar Orbit," *Apollo 11 Flight Journal*, https://history.nasa.gov/afj/ap11fj/11day4-loi1.html, 075:59:08–075:59:15.

51. W. David Woods, Kenneth D. MacTaggart, and Frank O'Brien, eds., "Day 5: Undocking and the Descent Orbit," *Apollo 11 Flight Journal*, https://history.nasa.gov/afj/ap11fj/15day5-undock-doi.html, 099:46:21.

52. Ibid., 100:37:31–100:38:56.

53. Eric M. Jones, ed., "The First Lunar Landing," *Apollo 11 Lunar Surface Journal*, https://www.hq.nasa.gov/alsj/a11/a11.landing.html, 102:38:26–102:38:42.

54. Ibid., 102:44:04–102:44:26.

55. Armstrong, interview by Ambrose and Brinkley, p. 83.

56. Neil Armstrong, commentary, in "First Lunar Landing," *Apollo 11 Lunar Surface Journal*, at 102:45:32.

57. Armstrong, interview by Ambrose and Brinkley, p. 86.

58. Eric M. Jones, ed., "Post-Landing Activities," *Apollo 11 Lunar Surface Journal*, https://www.hq.nasa.gov/alsj/a11/a11.postland.html, 103:02:03.

59. Ibid., 102:56:02.

60. Ibid., 105:25:38.

61. Buzz Aldrin, interview by the author, October 2005.

62. Hansen, *First Man*, p. 489.

63. Eric M. Jones, ed., "One Small Step," *Apollo 11 Lunar Surface Journal*, https://www.hq.nasa.gov/alsj/a11/a11.step.html, 109:24:48.

64. Ibid., 109:32:26.

65. Ibid., 109:43:16–109:43:24.

66. Armstrong, interview by Ambrose and Brinkley, p. 84.

67.  Aldrin, interview by the author.

68.  Jones, "One Small Step," 109:52:40.

69.  Eric M. Jones, ed., "Mobility and Photography," *Apollo 11 Lunar Surface Journal*, https://www.hq.nasa.gov/alsj/a11/a11.mobility.html, 110:16:30.

70.  Ibid., 110:17:44.

71.  "The Apollo 11 Technical Crew Debriefing, July 31, 1969," *Apollo 11 Lunar Surface Journal*, https://www.hq.nasa.gov/alsj/a11/a11tcdb.html.

72.  Lee Silver, in "Mobility and Photography," ed. Jones, at 110:27:20.

73.  Aldrin, interview by the author.

74.  Armstrong, interview by Ambrose and Brinkley, p. 74.

75.  Eric M. Jones, ed., "EASEP Deployment and Closeout," *Apollo 11 Lunar Surface Journal*, https://www.hq.nasa.gov/alsj/a11/a11.clsout.html, 111:15:13.

76.  "Apollo 11 Technical Crew Debriefing."

77.  "Apollo 11 Post Flight Press Conference, 16 September 1969," transcript, part 3, *Apollo 11 Lunar Surface Journal*, https://history.nasa.gov/ap11ann/FirstLunarLanding/ch-4.html.

78.  Eric M. Jones, ed., "The Return to Orbit," *Apollo 11 Lunar Surface Journal*, https://www.hq.nasa.gov/alsj/a11/a11.launch.html, 124:21:54.

79.  W. David Woods, Kenneth D. MacTaggart, and Frank O'Brien, eds., "Day 6: Rendezvous and Docking," *Apollo 11 Flight Journal*, https://history.nasa.gov/afj/ap11fj/19day6-rendezvs-dock.html, at 127:50:11.

80.  W. David Woods, Kenneth D. MacTaggart, and Frank O'Brien, eds., "Day 6: Boarding Columbia and LM Jettison," *Apollo 11 Flight Journal*, https://history.nasa.gov/afj/ap11fj/20day6-reboard-lmjett.html, 130:11:05.

81.  David M. Harland, *The First Men on the Moon: The Story of Apollo 11* (Berlin: Springer-Praxis, 2007), p. 311.

82.  W. David Woods, Kenneth D. MacTaggart, and Frank O'Brien, eds., "Day 7: Leaving the Lunar Sphere of Influence," *Apollo 11 Flight Journal*, https://history.nasa.gov/afj/ap11fj/22day7-leave-lsi.html, 151:39:56.

83.  Hansen, *First Man*, p. 1.

84.  Deborah Rieselman, "Neil Armstrong: 'Mr. Average Guy,'" *UC Magazine*, https://magazine.uc.edu/editors_picks/recent_features/armstrong/average.html.

85.  Deborah Rieselman, "Farewell Neil Armstrong: UC's Most Famous,

Humble Professor," *UC Magazine*, April 2013, https://magazine.uc.edu/issues/0413/Armstrong.html.

86.  Hansen, *First Man*, p. 602.

87.  Armstrong, interview by Ambrose and Brinkley, p. 92.

88.  "Joint Meeting of the Two Houses of Congress to Receive the Apollo 11 Astronauts," Congressional Record, September 16, 1969, https://www.hq.nasa.gov/alsj/a11/A11CongressJOD.html.

89.  Armstrong, interview by Ambrose and Brinkley, p. 79.

90.  "Joint Meeting of the Two Houses of Congress."

## CHAPTER 7. PETE CONRAD: SALTY SAILOR OF THE SKIES

1.  Tom Wolfe, *The Right Stuff* (New York: Farrar, Straus and Giroux, 1979), p. 71.

2.  Nancy Conrad and Howard A. Klausner, *Rocketman: Astronaut Pete Conrad's Incredible Ride to the Moon and Beyond* (New York: New American Library, 2005), p. 116.

3.  Wolfe, *Right Stuff*, p. 84.

4.  Conrad and Klausner, *Rocketman*, p. 139.

5.  Mark Wade, "Gemini 5," *Encyclopedia Astronautica*, http://www.astronautix.com/g/gemini5.html.

6.  Conrad and Klausner, *Rocketman*, p. 146.

7.  Susan Howlett, "Lunar Rover," *Los Angeles Times*, February 11, 1996, http://articles.latimes.com/1996-02-11/news/ls-35175_1_lunar-surface.

8.  Conrad and Klausner, *Rocketman*, p. 140.

9.  Ibid., p. 148.

10.  Ibid., p. 149.

11.  Armstrong swore to the end that he intended to say "a man," and thought he did. The audio recordings of the era seem to indicate otherwise, but in any case, that was his intention.

12.  Conrad and Klausner, *Rocketman*, p. 175.

13.  This and all the following quotes from the flight and lunar surface are from NASA's *Apollo 12 Lunar Surface Journal*, https://www.hq.nasa.gov/alsj/a12/a12.html.